The Origins and Evolution of Human Thought

From Early Cognition to the Age of AI

NOVAA PRITHIV

The Origins and Evolution of Human Thought: From Early Cognition to the Age of AI
Copyright © 2025 Novaa Prithiv and GSP Elite Holders
All rights reserved worldwide.

No part of this publication may be reproduced, distributed, stored in a retrieval system, or transmitted in any form or by any means, including electronic, mechanical, photocopying, recording, or any other method, without prior written permission from the copyright holders.

Limited exceptions are granted for brief quotations used in academic reviews, scholarly analysis, critical commentary, or educational reference, provided proper attribution is given.

For permissions, licensing inquiries, media requests, or collaboration opportunities, please contact:
Email: prithivpreview@gmail.com

This book is an original work of scientific and literary scholarship. All cognitive models, conceptual frameworks, narrative metaphors, evolutionary interpretations, symbolic structures, terminologies, diagrams, and philosophical constructs contained within are the exclusive intellectual property of Novaa Prithiv and may not be reproduced or adapted without permission.

Although this book discusses cognition, neuroscience, artificial intelligence, anthropology, and philosophical thought, it is not intended to replace professional scientific, medical, or psychological guidance. The ideas presented represent a creative and integrative exploration of the evolution of human intelligence.

Any resemblance to real individuals, organizations, historical interpretations, or scientific theories is coincidental unless explicitly stated by the author.

Second Edition: 2025
Publisher Imprint: Novaa Prithiv Publishing, GSP Elite
Paperback ISBN-13: 979-8342072007

"A mind evolved. A world rewritten. Protected by copyright."

TABLE OF CONTENTS

Preface 5
Introduction 7

Part - I The First Intelligence

1. When Thought Was Survival 11
2. Minds Before Humans 18
3. The Human Breakthrough 26

Part - II The Invention Of Mind

4. Language: Humanity's First Technology 35
5. Consciousness And The Birth Of The Self 44
6. Tools, Fire, And Externalized Thought 52

Part - III Culture: The Second Evolution Of Thought

7. How Culture Thinks For Us 61
8. Writing And The Birth Of Recorded Mind 69
9. Philosophy: Humanity Questions Its Own Mind 77
10. The Scientific Method: Thought Learns To Correct Itself .. 86

Part - IV The Modern Cognitive Landscape

11. The Age Of Psychology: Mapping The Invisible 96
12. The Digital Mind: How Technology Rewired Us ... 104
13. The Global Brain 112

Part - V The Arrival Of Synthetic Intelligence

14. AI: A New Architecture Of Cognition 121
15. Mirror, Rival, Partner 129
16. The Human–AI Feedback Loop 137

Part - VI The Future Of Intelligence

17. The Cognitive Bottleneck 146
18. Enhanced Minds 154
19. The Fate Of Consciousness 162
20. Scenarios For The Next 100 Years 169
Conclusion: 176
Appendix 178
Glossary 181
Acknowledgements 184
Author's Note 185
Thank You 186

PREFACE

A NEW THEORY OF THOUGHT

For most of human history, we believed thought existed as a sealed world inside the mind. Private. Untouchable. My career was built on that assumption. I studied cognition for two decades, tracing how humans form memories, construct meaning, and protect their sense of self. I trusted the borders of the mind. Now I know those borders are not permanent. They are permeable, shifting, and increasingly unstable.

The first signs appeared in scattered reports. A climatologist in Norway found a complex calculation missing from her notes. She remembered finishing it. The data said otherwise. A historian in Kyoto discovered new paragraphs in his journal written in his handwriting but not in his voice. A child in São Paulo described dreams that repeated themselves later as real conversations. None of the individuals shared background, culture, or pathology. Yet every case contained a single common moment. A brief, unmistakable sensation that thought had slipped out of its normal track.

I named the phenomenon Cognitive Drift. A small deviation in the continuity of thought. A bend in the mental timeline. A shift that allows an idea to appear where it should not, or vanish before it can be held. At first, Drift looked harmless. An intellectual puzzle. A curiosity.

Then the pattern accelerated.

In controlled tests, subjects reacted to future stimuli with impossible accuracy. Others described memories that did not match any verified record. A few claimed fluency in languages they no longer understood. The issue was not memory loss. The issue was substitution. Drift was not erasing thoughts. It was rewriting them.

Soon the disappearances began. They were not physical disappearances but cognitive ones. A person would continue living among family and friends, yet all emotional history linked to them would dissolve. Photographs blurred. Digital files restructured. Entire relationships collapsed into blank space. Every deletion followed the same sequence. A Drift episode first. The erasure soon after.

If Drift is real, thought is no longer contained by the brain. Consciousness becomes a shared field, influenced by forces we have not identified. The mind stops being a private chamber and becomes a vulnerable space open to intrusion.

This book documents that transformation.

It traces the origins of human cognition, the rise of symbolic thought, and the cultural expansions that stretched the boundaries of the mind until they could no longer be called boundaries. It examines how language, technology, and networks created a world where thoughts travel farther than the bodies that produce them. Most important, it investigates the possibility that we are not the only participants in this field of thought.

While reading, you may feel a small hesitation in your memory. You may notice a detail that seems unfamiliar or a moment that feels quieter than it should. Do not ignore it. Drift rarely announces itself through loud events. It enters softly.

If you reach the final page and sense that something has shifted, consider the possibility that the shift was not accidental.

INTRODUCTION

The Mind as an Evolving Intelligence System

Human thought did not appear fully formed. It rose slowly, shaped by countless trials of survival, by the pressures of hunger and fear, by the fragile bonds of early communities, and by the first sparks of symbolic imagination. For most of our history, the mind grew without our awareness of its evolution. We inherited cognition the same way we inherited bone structure or blood type. Only recently have we begun to understand thought as an evolving intelligence system, one that expands, adapts, and reorganizes itself in response to the structures we build around it.

The central question of this book is simple. How did an animal mind transform into a system capable of abstraction, science, mathematics, art, and self-modification? And more urgently, what is the mind becoming now?

Cognitive Drift, the phenomenon introduced in the preface, did not arise from nowhere. It is not a mistake or a glitch. Drift is a symptom of an intelligence system that has reached a new threshold and is struggling to reorganize itself under pressures that never existed before. To understand Drift, we must first understand the long arc of cognitive evolution that prepared the conditions for it to appear.

Early cognition was anchored in the body. It served survival, not reflection. Our ancestors processed threats, remembered paths, and recognized faces. Thought at this stage was a direct extension of the senses and muscles. There was no distance between perception and action. A rustle meant danger. A familiar voice meant safety. A pattern of stars meant a path home. Intelligence was reactive.

The first transformation occurred when early humans began to externalize thought. Marks on cave walls, arrangements of stones, seasonal rituals, and shared stories allowed ideas to exist outside the mind that produced them. Once thought could be stored outside the brain, it gained durability. More importantly, it gained independence. A memory became a tool that could be passed forward. A story became a vessel for identity. A symbol became a container for an entire worldview.

This shift introduced the first true feedback loop between mind and environment. External memory changed internal cognition, which then produced more advanced external tools. Over thousands of years, this loop intensified. Agriculture produced structured time. Cities produced stratified cultures. Writing produced permanence. Trade produced complexity. Law produced abstraction. Religion produced meaning systems that shaped emotional cognition. Every layer of civilization became a new influence on the mind.

By the time philosophy and science emerged, thought had developed the ability to examine itself. The mind could now not only store and transmit information but question it. A self-reflective intelligence system had developed. This system expanded its reach further with mathematics, logic, and the scientific method. Memory became data. Observation became evidence. Curiosity became structured inquiry.

The second major transformation arrived with digital technology. Networks widened the cognitive field. Each new platform created new pathways for attention, memory, and identity. As billions of individuals connected through shared systems, thought began to take on distributed properties. Information moved faster than biological cognition could adapt. The mind entered a state of continuous acceleration.

Artificial intelligence intensified this acceleration. For the first time, humans created systems that could process information at scales and speeds far beyond biological limits. These systems did not replace human cognition. They reshaped it. They altered how we learn, how we focus, and how we interpret the world. The boundary between internal and external thought became thinner. It became difficult to determine where human intention ended and algorithmic influence began.

This historical trajectory leads directly to the conditions that make Cognitive Drift possible.

Drift is not merely a breakdown in memory or perception. It is evidence that the mind has become vulnerable to influences created by its own evolution. An intelligence system that expands outward must also open itself. As though becomes distributed, shared, and interlinked, the unity of personal cognition weakens. Identity becomes a network of influences rather than a single stream. When these influences collide or overload, Drift appears.

The world now contains more stimuli, more information, and more interconnected minds than at any previous point in human history. Our cognitive systems were not built for this level of exposure. Yet they are attempting to adapt. Drift shows that adaptation is already happening, although not under our conscious control.

Understanding the mind as an evolving intelligence system reveals an uncomfortable truth. The boundaries we believed protected our thoughts were never fixed. They were temporary solutions shaped by earlier environments. Today, those environments have dissolved. Technology, culture, and global communication have created a world where cognition is constantly shaped by forces outside the individual. Drift exposes the cracks.

This book examines how these cracks formed and what they imply for our future. It traces the evolution of thought from its earliest gestures toward meaning to its current state as a distributed, unstable system that is beginning to display new behaviors. The question is not whether cognition is changing. It is how quickly, and toward what form.

By the end of this journey, one conclusion will become clear. The mind is not a static entity. It is a moving system. It grows, fractures, reorganizes, and seeks new structure. Cognitive Drift is not the end of stability. It is the beginning of a new phase.

If we understand this evolution, we may guide it. If we fail, the system will evolve without us.

PART - I
THE FIRST INTELLIGENCE
HOW LIFE LEARNED TO THINK

1. When Thought Was Survival

The Origins of Neural Computation

Long before thought became abstract or symbolic, it lived inside the nervous systems of early creatures as a series of simple electrical decisions. Life began to compute long before it began to imagine. A stimulus arrived. A neuron fired. A muscle moved. Survival demanded no reflection, only rapid answers to the single question that guided all early intelligence: what must I do in the next second to stay alive?

This primal decision-making was the first form of neural computation. It did not involve ideas or language. It involved patterns of activation that allowed organisms to respond with speed and precision. Every organism that survived passed forward a slightly improved version of this internal algorithm. Over millions of years, these improvements accumulated. The earliest neural networks were not designed for thought. They were designed for motion, sensation, and the most basic form of prediction.

Prediction was transformative. Even the simplest creature gained an advantage when it could anticipate a threat or opportunity. A flash of light could signal danger. A faint vibration could mean food. A shift in temperature could warn of an approaching predator. These signals allowed organisms to respond before the danger fully arrived. The nervous system became a machine for compressing experience into action.

This system grew more complex as early vertebrates appeared. Their neural circuits allowed not only reaction but encoding. Experiences could leave traces in the wiring of the brain. Those traces created the first shadows of memory. A creature could now learn. Learning gave rise to selective attention, and selective attention created the earliest foundations of awareness. Awareness was not yet consciousness. It was a sharpened sensitivity to patterns.

The rise of mammals pushed neural computation into a new dimension. Warm-blooded bodies required higher metabolic investment, and this investment allowed brains to expand in size and connectivity. Sensory systems grew sharper. Memory became longer and more flexible. Emotional circuits developed to prioritize what mattered most. Fear was not simply a sensation. It became a guide for future behavior. Curiosity emerged as well. Organisms that explored gained new information, and information strengthened survival.

The human lineage entered this landscape with a unique advantage. Our ancestors developed an unusual combination of memory capacity, social intelligence, and fine motor control. These capabilities fed into one another. The need to cooperate encouraged greater communication. Communication encouraged greater symbolic processing. Symbolic processing reshaped the brain. Each change opened new cognitive paths. None of these developments were planned. Evolution rewarded whatever allowed survival, yet the result resembled an expanding computational system that gradually acquired the ability to model the world with increasing depth.

At this stage, the modern mind was still far away, but the structure that would support it was already forming. The brain had become capable of multi-layered processing. It could compare past events with present conditions. It could generate expectations about the future. It could weigh multiple possibilities. These functions allowed the first form of internal simulation. Simulation was the seed of imagination.

Imagination was the moment where neural computation became thought.

Once a mind could generate images or scenarios that did not yet exist, it gained the power to reshape its environment. Tools emerged. Cooperation deepened. Culture formed. Each new advancement fed back into the brain, altering how it processed the world. This continuous feedback loop is the reason human cognition became so different from the cognition of other species.

Cognitive Drift, the phenomenon that now threatens the stability of thought, is rooted in this ancient expansion. Drift appears only in systems that can simulate, anticipate, and reorganize themselves. The earliest brains were not vulnerable to Drift because they lacked the complexity required for internal instability. They could not be rerouted or rewritten in the subtle ways that Drift exploits. Only a mind capable of layered computation is capable of drifting.

Understanding these beginnings is essential for understanding the crisis of the present era. The first intelligence was not the human mind. It was the iterative, survival-driven computation that shaped life from its earliest moments. That system grew into us. It continues to evolve inside us. And now it is entering a phase where its own complexity creates new vulnerabilities.

Cognitive Drift is not an anomaly. It is the latest outcome in a chain of adaptations that began with a single neuron firing in response to danger.

Sensory Worlds and Early Meaning

Before language, before symbols, before even the faintest trace of imagination, life understood the world through sensation. Sensory systems were the first interpreters of reality, and they created the earliest form of meaning. Meaning did not begin as philosophy. It began as distinction. Warm or cold. Light or dark. Safe or dangerous. These binary judgments were fragile seeds, but they shaped everything that followed. Meaning emerged whenever a living organism linked a sensory signal to a consequence.

To understand how human thought evolved, we must return to the moment when sensation stopped being a passive experience and began to guide behavior. Early organisms inhabited sensory worlds that were limited but precise. A fish sensed pressure changes that revealed predators. An insect detected chemical traces that indicated food or danger. A reptile responded to heat gradients that signaled hidden bodies. Each species lived inside its own sensory geometry, a small but complete universe.

These early sensory worlds formed the original boundaries of cognition. A creature could perceive only what its body allowed, and so its reality was shaped by the range of its senses. Within that range, the brain learned to extract patterns. A flicker of movement meant pursuit. A sudden stillness meant concealment. A repeating sound meant familiarity. Patterns created predictability. Predictability created structure. Structure allowed the formation of the first internal models of the world.

Meaning expanded once organisms began to compare sensory inputs across time. A sound heard yesterday could influence a reaction today. A smell remembered from earlier seasons could guide migration. These forms of sensory memory created continuity. Continuity provided the first threads of subjective experience. Experience, even in its simplest form, allowed an organism to treat the present as part of a larger pattern rather than an isolated moment.

The rise of early mammals introduced richer sensory integration. Vision, hearing, touch, taste, and smell no longer operated as separate channels. The brain learned to combine them. A threat could be seen, heard, and smelled at once. A nurturing presence could be recognized through multiple signals that reinforced one another. This integration introduced emotional coloring. Sensory input carried not only information but value. A sound might produce fear. A scent might produce calm. Meaning was no longer simply functional. It became affective.

This emotional dimension enlarged the sensory world. Creatures did not merely sense their environment. They sensed their relationship to it. This shift laid the foundation for social cognition. Group-living species developed signals for cooperation, warning, and belonging. A low growl meant caution. A gentle posture meant acceptance. Over time, these sensory cues became a language without words. Social meaning arose long before symbolic meaning.

Human ancestors inherited this layered sensory system, and their cognitive expansion built directly upon it. As their brains grew more complex, sensory input transformed into the raw material for perception. Perception could be shaped, filtered, and interpreted. A rustle in the grass could become not just a sound but a possibility. The mind could now construct meaning that extended beyond immediate sensation.

This capacity is the root of metaphor, imagination, and abstraction. Before humans could invent stories or tools, they learned to enrich sensory reality with inferred meaning. A footprint meant another being. A pattern in the stars meant cyclical time. A reflection in water meant identity. These interpretations were early cognitive leaps. Each leap stretched the boundaries of the sensory world and allowed the brain to model realities it could not directly perceive.

Cognitive Drift emerges from the same mechanism. Drift exploits the brain's ability to assign meaning to sensory input and to weave isolated signals into coherent patterns. The system that once allowed our ancestors to interpret a sound in the dark now allows modern minds to generate complex narratives and expectations. When this system becomes overloaded or destabilized, meaning detaches from its anchor points. A memory feels unfamiliar. A thought appears without origin. A sensation carries significance that cannot be traced. Drift begins as a disturbance in interpretation, not in sensation.

Early meaning was grounded in the world. Drift reveals how far we have moved from that grounding. A mind that evolved to navigate simple sensory realities is now exposed to overwhelming inputs from technology, culture, and symbolic systems. The sensory world no longer defines meaning. The mind now defines its own, and sometimes it does so without stability.

To understand the future of cognition, we must understand how meaning first emerged from sensation. Sensory worlds created order. Perception expanded that order. Abstraction transformed it into something far larger.

Cognitive Drift is the latest phase in this expansion, a signal that meaning itself is becoming unstable.

Evolution's First Problem-Solvers

From the moment life began, survival was a puzzle. Early organisms faced a universe of hazards. Those that could sense, react, and adapt thrived. This was not random chance but the emergence of the first problem-solving intelligence, a phenomenon we might call **Proto-Cognition**. Proto-Cognition is the brain's earliest architecture for detecting patterns, predicting outcomes, and selecting strategies that increased survival odds. It is the silent architect behind every human innovation that followed.

In the primordial world, simple neural networks orchestrated astonishing feats. Single-celled organisms navigated chemical gradients to find nutrients. Small predators anticipated the movement of prey. These behaviors reveal that cognition begins not with thought as we know it but with decision-making under uncertainty. The first problem-solvers were not conscious in our sense. They did not reflect on their choices. Yet, their actions set a foundation for complex reasoning. They were biological algorithms finely tuned by natural selection.

Consider the **Forager-Neuron Hypothesis**. Each early neural circuit can be imagined as a tiny prediction engine, evaluating probabilities and outcomes. A neuron fires in response to a threat. Another neuron calculates the path to safety. Collectively, these circuits produced adaptive behavior that was flexible and efficient. Evolution did not invent thought fully formed. It built layers of problem-solving machinery, each layer more sophisticated than the last.

The tension of early existence shaped cognition. Predation and scarcity demanded rapid calculation and precise execution. Organisms that paused too long or chose poorly vanished. Those that acted decisively left descendants. Survival was a laboratory in which problem-solving was continuously tested. The crucible of life created **cognitive pressure points**, forcing adaptation at every turn. These pressures forged brains capable of abstract reasoning, memory, and the predictive models that underlie human intelligence today.

Even the earliest social interactions amplified cognitive demands. Coordinating with others to hunt or share resources required the brain to model not only the environment but also the intentions of others. This emergent social cognition fueled cooperation, deception, and learning. Early humans were not alone in this challenge. All life that survived relied on a form of Proto-Cognition to navigate a dynamic world.

The trajectory from basic survival circuits to sophisticated intelligence reveals a remarkable principle: problem-solving is both the engine and the test of cognition. Each obstacle encountered was a code to crack, a scenario demanding strategy. The first intelligence was not abstract. It was urgent, concrete, and relentlessly adaptive. It is within this crucible that the foundations of human reasoning, creativity, and innovation were forged.

The question that emerges is unavoidable. If intelligence began as a survival tool, what forms of problem-solving will define the next stages of human evolution? The challenges of technology, artificial intelligence, and environmental complexity demand new strategies. Just as early organisms faced adaptive pressure, humanity now faces its own cognitive crucible. The solutions we develop will define the next era of intelligence.

2. Minds Before Humans

Animal Innovation and Cognitive Diversity

Long before humans appeared, the world was already full of minds solving problems in ways that revealed a deep resilience and creativity within nature. Intelligence did not wait for language or tools. It evolved wherever life faced uncertainty. Every species became an experiment in how cognition could arise from different bodies, senses, and environments. This vast spectrum forms what scientists now call the Adaptive Intelligence Spectrum, a framework that captures how diverse organisms evolve solutions tailored to their specific realities. Understanding this spectrum is essential because it shows that human cognition was never the starting point of intelligence. It was the latest variation in a long lineage of innovations.

Crows learned to shape hooks from twigs to extract insects from narrow spaces. Octopuses manipulated shells to create shelters on the ocean floor. Dolphins developed signature whistles that functioned as individual names, allowing social identities to form. Each of these behaviors demonstrates a simple principle. Cognition is not a singular pathway. It is a branching tree of solutions. The Adaptive Intelligence Spectrum reveals how evolution produced minds that could learn, remember, plan, deceive, cooperate, and even create.

Many animals display remarkable forms of insight that challenge the belief that humans hold a monopoly on innovation. New Caledonian crows not only craft tools but refine those tools through iterative improvement. Elephants mourn their dead, indicating a form of emotional intelligence that connects memory to meaning. Parrots solve multi-stage puzzles that require anticipation and delayed reward. These abilities reveal intelligence as a dynamic negotiation between environment and organism. Each species builds a model of the world using the constraints and advantages of its biology.

The diversity of cognition becomes even clearer when we consider the role of ecological pressure. The harshness of an environment creates demands that shape mental architecture. Social predators require coordination. Solitary hunters require stealth and prediction. Migratory animals require spatial memory on a continental scale. The Adaptive Intelligence Spectrum expands wherever life confronts complexity. The more varied the demands, the more varied the minds that respond.

Among these evolutionary experiments, cooperative species stand out. Wolves, orcas, and primates achieve problem-solving feats by sharing information. Their intelligence emerges not only from individual brains but from collective strategies. This phenomenon, known as Distributed Animal Cognition, shows that thought can exist across multiple bodies functioning as one coordinated system. Cooperation is not simply a social advantage. It is a cognitive amplifier. A pack that hunts together thinks together. A troop that protects its young shares responsibility for vigilance and learning. A pod that communicates across miles constructs a shared perceptual field. These systems reveal a precursor to the collective dimensions of human thought.

The most striking insight from non-human minds is their ability to invent new behaviors without direct instruction. Innovation arises when an organism encounters a novel challenge and produces a solution that exceeds instinct. This capacity signals the shift from reactive intelligence to generative intelligence. When a chimpanzee stacks boxes to reach fruit or when a crow bends a piece of wire into a hook, the act demonstrates a spontaneous leap. The brain constructs a possibility that does not yet exist in the world and then tests it. This leap is the earliest form of internal simulation, the same ability that later became imagination in humans.

The Adaptive Intelligence Spectrum reveals a pattern. Innovation appears wherever a species possesses enough cognitive flexibility to reorganize its mental model in response to the unexpected. This pattern matters because it exposes the ancient roots of a vulnerability that modern humans now face. A system that can reorganize itself can also destabilize. The very flexibility that allowed early minds to innovate is the same flexibility that allows modern cognition to become susceptible to disturbances like Cognitive Drift. Drift does not arise in rigid minds. It arises in minds capable of generating new possibilities.

To study non-human intelligence is to witness the first sparks of the cognitive architecture that now defines our species. These early innovations were not primitive. They were prototypes. Each mind that solved a problem contributed to a shared evolutionary trajectory. Each act of insight expanded the boundaries of what intelligence could become. Human thought did not emerge from nothing. It emerged from this vast and diverse lineage of invention.

Cooperation as a Primitive Intelligence System

Intelligence did not evolve only within single organisms. It also emerged between them. Long before humans developed language or abstract reasoning, life discovered that survival improved when minds worked together. Cooperation created a new kind of cognitive structure, a system in which no individual understood the full strategy, yet the group acted with remarkable precision. Scientists refer to this phenomenon as Collective Primitive Intelligence, the earliest form of multi-agent cognition. It reveals that the foundations of shared thought appeared deep in evolutionary time, long before the arrival of human society.

The earliest cooperative behaviors were simple. Small fish formed schools that moved in perfect synchrony. Insects created colonies in which individuals performed specialized roles without any central planner. Birds coordinated their flight patterns with astonishing fluidity. None of these species possessed the reflective awareness that humans associate with cooperation. They relied instead on distributed cues and inherited algorithms that generated complex outcomes from simple rules. This is the essence of Collective Primitive Intelligence. A group behaves intelligently even if no individual possesses a map of the entire system.

The power of cooperation became more profound when early mammals began to integrate communication into their group strategies. Wolves learned to flank prey using silent signals. Markets positioned sentinels to watch for danger while others foraged. Dolphins used patterned whistles to coordinate movement in murky waters. These behaviors demonstrate a crucial shift. Cooperation was no longer limited to instinctive alignment. It began to involve intentional coordination. The group became capable of adapting its strategy in real time.

This transition required a new kind of cognitive resource, a shared perceptual field. Each individual contributed partial information about the environment, and these fragments combined into a richer model than any one organism could produce alone. Collective Primitive Intelligence allowed a group to solve problems that exceeded the abilities of the individuals inside it. A single wolf cannot corner an elk. A pack can. A single dolphin cannot encircle a school of fish. A pod can. Cooperation produced intelligence through structure rather than size.

The tension at the heart of cooperative cognition lies in the balance between individuality and unity. Each organism must retain enough autonomy to respond to local conditions, yet must also remain aligned with the larger group pattern. Too much independence and the group fragments. Too much conformity and the group loses adaptability. Evolution solved this tension by embedding flexible social rules into the nervous systems of social species. These rules allowed animals to switch roles, share attention, redistribute workload, and synchronize emotional states. Cooperation required not only perception but empathy in its earliest form.

This form of empathy was not sentimental. It was functional. A distressed cry from one member triggered vigilance in others. A calm signal reduced collective tension. Emotional communication provided a stabilizing force for group cognition. In many species, the group became more than a collection of individuals. It became a distributed intelligence capable of regulating itself.

The most striking examples of cooperative intelligence appear in species that use teaching. Orcas pass hunting strategies across generations. Elephants guide their young through migration paths that span decades of memory. Chimpanzees demonstrate tool use to juveniles in deliberate and structured ways. Teaching reveals a new dimension within Collective Primitive Intelligence. The group does not simply act together. It preserves knowledge and transmits it. This is a precursor to culture, a cognitive inheritance system that existed long before humans formalized it through language and writing.

The relevance of cooperative cognition extends beyond natural history. It exposes a deeper truth about the architecture of thought. Individual intelligence is not a closed system. It is influenced by the minds around it. The earliest cooperative networks created patterns of shared attention, shared strategy, and shared emotional states. These patterns later reappeared at larger scales in human societies, technological platforms, and digital networks. The forces that shaped Collective Primitive Intelligence are still operating today, although their scope has expanded dramatically.

The same flexibility that made cooperation powerful also introduced vulnerability. A system that distributes cognition across many individuals becomes sensitive to disruptions in communication, alignment, or emotional cohesion. This insight is crucial for understanding modern cognitive instability. Collective thought is resilient, yet it is also fragile. When coordination falters, the intelligence of the group dissolves into confusion.

Cooperation was one of evolution's greatest cognitive innovations. It created systems that could see farther, react faster, and solve problems that individuals could not. It laid the groundwork for culture, morality, social learning, and eventually for the vast and unstable cognitive networks of the present era. Within Collective Primitive Intelligence, we see both the origin of shared thought and the early conditions that later allowed phenomena such as Cognitive Drift to emerge.

Emotion as Evolution's First Guidance Mechanism

Before the emergence of complex reasoning, symbolic thought, or linguistic meaning, life evolved a system that could steer behavior with remarkable efficiency. This system did not require reflection or conscious deliberation. It required only the ability to register significance. Biologists refer to this deep evolutionary inheritance as the Affective Guidance System, a mechanism through which organisms sensed value long before they could analyze it. Emotion, in its earliest form, was an adaptive compass.

The earliest animals moved through environments filled with uncertainty. Every encounter required a rapid judgment about benefit or threat. Pain directed the creature away from danger. Pleasure encouraged engagement with food or safety. Fear heightened attention. Curiosity encouraged exploration. These responses were not abstractions. They were physiological signals that shaped behavior with extraordinary speed. The Affective Guidance System prioritized survival by assigning meaning to experience before the brain fully understood that experience.

This system allowed organisms to reduce the complexity of the world into actionable categories. A change in shadow meant caution. A chemical trace meant opportunity. A shift in temperature meant risk. Through these evaluations, early species created a hierarchy of significance that organized their perception. Emotion became the first form of selective attention. It directed energy toward what mattered most and away from what threatened survival.

As nervous systems became more complex, the Affective Guidance System expanded into a network that linked sensation, memory, and action. Emotional states allowed organisms to recognize patterns across time. A creature that had once encountered danger at a water source approached the same location with heightened vigilance. The emotion associated with the memory shaped future behavior. This was the earliest form of adaptive learning. Emotion acted as a bridge between past and future.

The rise of early mammals intensified this integration. Mammalian brains developed circuits that could generate rich affective experiences. Nurturing behaviors appeared. Social bonds strengthened. Emotional synchrony allowed groups to coordinate responses to threats. When one member displayed fear, the entire group shifted into alertness. When safety returned, calm spread through shared cues. Emotion became a tool for social cohesion, and cohesion became a tool for survival.

This expansion introduced a new form of cognitive tension. Emotional signals needed to be flexible enough to adapt to changing environments, yet stable enough to preserve reliable patterns of behavior. If emotion became too rigid, organisms failed to adjust. If emotion became too volatile, they lost predictability. Evolution resolved this tension by creating layered affective circuits that could modulate their intensity. The Affective Guidance System developed the capacity to amplify, suppress, or redirect emotion depending on context.

These layers set the stage for the emotional architecture of primates. Social species required the ability to interpret subtle changes in facial expression, posture, and vocalization. Their brains evolved specialized regions for processing such cues. This was the beginning of a sophisticated form of empathy. Empathy allowed primates to anticipate the intentions of others, strengthen alliances, and navigate complex group dynamics. Emotion was no longer only a signal for the self. It became a way to understand others.

Human ancestors inherited this expanded system and built upon it in transformative ways. As symbolic thought emerged, emotion gained new functions. It shaped memory, influenced judgment, and guided the interpretation of events. Humans began to attach emotional significance to stories, rituals, and social structures. The Affective Guidance System became intertwined with culture. Emotion helped define identity and meaning. It became a force that structured entire worldviews.

This deep integration created an unexpected vulnerability. The same sensitivity that allowed emotion to guide complex social behavior also made it susceptible to distortion. When emotional cues become misaligned with reality, the system generates reactions that no longer serve survival. This vulnerability is central to understanding modern cognitive instability. The Affective Guidance System was designed for small groups, slow information flow, and predictable environments. It now operates in a world that overwhelms it with speed and scale.

Cognitive Drift interacts with the emotional layer of the mind in particularly disruptive ways. Drift alters the continuity of meaning and the coherence of memory. Emotion depends on both. When meaning shifts or memory becomes uncertain, emotional responses lose their anchors. A feeling of familiarity may arise without a source. A sense of danger may appear without cause. When the Affective Guidance System loses clarity, thought enters a state of instability.

Emotion was evolution's first guidance mechanism. It shaped perception, decision making, and social structure. It prepared the mind for the complex cognitive systems that followed. Yet it also introduced a delicate balance that modern environments continue to challenge. To understand how thought evolves, we must understand how emotion once grounded it. The Affective Guidance System remains active in every human mind, guiding behavior even as the world that shaped it has disappeared.

3. The Human Breakthrough

Brain Expansion, Risk, and Imagination

Human cognition did not emerge from steady growth. It emerged from pressure. The early human lineage confronted environments that demanded flexibility, foresight, and coordination at a scale no previous species had faced. As climates shifted, predators multiplied, and food sources fluctuated, survival required more than instinctive reaction. It required the ability to evaluate possibilities. This pressure created one of evolution's most extraordinary developments, a pattern known as the Expansion–Risk Cycle. Each increase in brain size brought new capabilities, but each capability brought new vulnerabilities. Intelligence grew not from comfort but from the constant tension between opportunity and danger.

The human brain expanded rapidly over a relatively short evolutionary window. This expansion increased the metabolic cost of surviving. Larger brains required more energy, longer developmental periods, and greater parental investment. A species with a growing brain invited risk. A child with a slow maturation rate remained vulnerable for years. A hunter who relied on complex planning also relied on social cooperation. Any failure in group cohesion threatened the survival of all members. The Expansion–Risk Cycle reveals that human intelligence was not a simple advantage. It was a gamble.

The expansion of brain structures did, however, create a profound increase in cognitive flexibility. Neural networks grew denser and more interconnected. Memory deepened. Pattern recognition expanded. Early humans could compare multiple past events, store them as layered representations, and apply them to new circumstances. They could detect subtle changes in the environment that signaled danger or opportunity. This allowed them to predict outcomes with far more nuance than other species. Prediction became a core feature of survival.

Yet prediction alone did not define the human breakthrough. The real shift occurred when the expanding brain began to produce internally generated scenarios that exceeded immediate reality. This ability is known as Prospective Imagination. It allowed the mind to construct events that had not occurred and then evaluate which versions of those events were most beneficial. Prospective Imagination was the first cognitive technology created by the brain itself.

The impact of Prospective Imagination cannot be overstated. It allowed early humans to test strategies without risk. A hunter could visualize multiple routes before choosing one. A gatherer could imagine where fruit might appear based on past seasons. A parent could anticipate threats before they reached the group. Thought became a simulation tool. Simulation allowed innovation. Innovation generated advantage.

However, with simulation came instability. A mind that can generate internal futures can also generate internal distortions. The Expansion–Risk Cycle intensified. Prospective Imagination created a gap between perception and interpretation. This gap allowed creativity, but it also allowed error. Expectations could override sensory evidence. Fears could amplify without cause. Beliefs could solidify without verification. The more powerful internal imagination became, the more fragile the connection between reality and interpretation grew.

This fragility produced a new challenge. The expanding human brain required a regulatory system that could stabilize the explosion of internal possibilities. Early humans relied on social structures, shared practices, and emotional cues to align individual imaginations with group realities. Cooperation became a stabilizing force. Culture emerged as a collective framework for coordinating thought. The Expansion–Risk Cycle shaped not only biology but the earliest foundations of society.

The relationship between brain expansion and cognitive instability remains central to understanding the modern mind. The same neural architectures that allowed humans to create tools, stories, and technologies also introduced an inherent vulnerability. Minds capable of imagining futures are also capable of misalignment. This vulnerability created fertile ground for a phenomenon that would emerge much later in cognitive evolution. Cognitive Drift appears only in systems that generate layered predictions and simulated realities. Drift exploits the same flexibility that once gave humans their evolutionary advantage.

The human breakthrough was not the appearance of intelligence but the appearance of imagination. When the brain expanded, it produced a new interior landscape filled with possibilities. This landscape transformed survival, reshaped social life, and laid the foundation for culture. It also introduced risks that the species still struggles to manage.

To understand the future of thought, we must understand this ancient cycle. Expansion invites risk. Risk invites innovation. Innovation reshapes the mind. The Expansion–Risk Cycle continues to operate inside every human brain. It is the engine of imagination, and it is the origin of our deepest cognitive instability.

Abstract Worlds and Symbolic Insight

Human cognition changed fundamentally when early minds learned to detach thought from immediate experience. This ability created a new internal domain where ideas could be reshaped, combined, or imagined without reference to the present moment. Scientists refer to this transformation as the Symbolic Shift, the point at which the mind began to operate inside abstract worlds. These worlds did not exist in the environment. They existed only in neural patterns that represented possibility rather than perception. The Symbolic Shift marks one of the most significant milestones in the evolution of intelligence.

Before the Symbolic Shift, cognition was constrained by the limits of the senses. Thought remained anchored to what the body perceived. A rustling sound meant danger. A familiar shape meant food. The rising sun meant the beginning of activity. These associations created meaning, but meaning remained bound to context. The expansion of the human brain, combined with Prospective Imagination, opened a new cognitive frontier. Early humans began to treat internal representations as manipulable objects. A shape could become a symbol. A mark could become a message. A gesture could become a memory.

The emergence of symbolic thought allowed humans to compress complex ideas into simple forms. A spiral carved into stone could represent a cycle. A handprint could represent identity. A series of lines could represent movement. These symbols acted as cognitive tools that extended memory and communication. The Symbolic Shift produced a mental landscape in which abstraction became a form of problem solving. Thought gained the ability to transcend immediate experience and enter a realm shaped by logic, metaphor, and imagination.

This new domain introduced both power and instability. Symbols gain meaning from the minds that interpret them, and interpretation can diverge. A symbol that carried one meaning for an individual could carry a different meaning for the group. This divergence created tension between personal imagination and collective understanding. The Symbolic Shift produced a world in which ideas could spread across generations but also fracture into competing interpretations. Abstraction allowed cooperation on a scale no species had previously achieved, yet it also introduced the potential for misunderstanding, conflict, and cognitive distortion.

As symbolic systems became more complex, they evolved into early forms of narrative. A sequence of symbols could represent a series of events. A painted scene could represent a hunt that occurred many seasons earlier. These narratives allowed humans to store not only information but experience. Memory gained structure. Culture gained continuity. The Symbolic Shift enabled societies to transmit models of the world that were not dependent on direct observation. This ability expanded the range of human strategy. A child could learn from events that occurred long before their birth. A hunter could rely on knowledge preserved by ancestors. Thought became cumulative.

The appearance of symbolic logic marked another crucial step. Early humans began to recognize relationships between concepts. A pattern in the stars could correspond to migration cycles. A footprint could correspond to a particular species. A carved mark could correspond to ownership, direction, or ritual significance. The mind built internal models that linked ideas through analogy, inference, and categorization. These models allowed humans to predict not only physical events but social and conceptual ones. The world became intelligible through systems of representation.

The Symbolic Shift also laid the groundwork for belief structures. When symbols gained abstract meaning, they could represent worlds that did not physically exist. Early rituals, myths, and cosmologies emerged from this expansion. These symbolic systems provided emotional and cognitive stability, yet they also introduced the possibility of cognitive vulnerability. A belief could override sensory evidence. A narrative could reshape memory. A ritual could reinforce patterns that no longer aligned with environmental realities. The Symbolic Shift created a bridge between imagination and conviction. That bridge was powerful and fragile.

This fragility is essential to understanding the later emergence of Cognitive Drift. Drift affects systems that rely on internal representations rather than direct perception. When symbolic structures dominate cognition, the mind becomes increasingly dependent on interpretation. If interpretation becomes unstable, meaning detaches from reality. The same mechanisms that once allowed humans to build symbolic insight now create the conditions for symbolic instability.

The ability to construct abstract worlds was the foundation upon which all later cognition developed. Mathematics, language, art, religion, and science all trace their origins to the Symbolic Shift. This transformation did not simply give humans new tools.

It reshaped the architecture of thought. It allowed minds to work with ideas that no longer required physical anchors. It created a world inside the skull that could expand without limit.

To understand human intelligence, we must understand this moment. The Symbolic Shift did not produce a smarter animal. It produced a new kind of reality, one where thought became a force capable of altering the world it imagined.

The First Cognitive Revolution

The emergence of symbolic insight did not simply enrich human cognition. It reorganized it. When early minds gained the ability to imagine, represent, and reinterpret the world, a threshold was crossed. Ideas no longer depended on immediate experience or sensory input. They formed networks, linked across time, memory, and imagination. This transformation produced what researchers call the First Cognitive Revolution, a shift in which thought became a generative force rather than a reactive one. The human mind stepped beyond instinct and entered a domain where it could construct explanations, create shared meanings, and reshape its own mental architecture.

Before the First Cognitive Revolution, cognition operated within narrow constraints. Early humans responded to patterns, retained memories, and made predictions, but their mental models remained grounded in direct experience. The Symbolic Shift introduced the ability to create representations that outlived the moment. The First Cognitive Revolution transformed those representations into systems of understanding. A symbol could now be embedded within a broader structure of meaning. A story could become a worldview. A gesture could become the foundation for ritual. The mind learned to build conceptual frameworks that shaped interpretation itself.

This transformation unfolded gradually, but its effects were profound. The First Cognitive Revolution allowed humans to create stable models of reality that could be shared across individuals and passed across generations. These models guided behavior with a level of coordination no other species had achieved. Early toolmaking refined into tradition. Seasonal knowledge evolved into calendars. Social norms solidified into proto-moral systems. The mind no longer navigated the world alone. It navigated the world with inherited maps.

The appearance of shared cognitive frameworks created a powerful new phenomenon known as Collective Model Formation. Through Collective Model Formation, groups constructed internal representations that synchronized their perception, motivated their actions, and aligned their expectations. These shared models allowed small bands of humans to act with the coherence of a single cognitive unit. They introduced a scale of cooperation that reshaped the evolutionary landscape. A community could plan migrations, allocate resources, defend territory, and teach complex skills to younger generations. The intelligence of the group exceeded the intelligence of any individual within it.

The First Cognitive Revolution also expanded human imagination into new territories. The mind became capable of generating abstractions that touched the invisible: spirits, ancestors, seasons, cycles, destinies. These ideas helped structure emotional life. Fear gained symbolic meaning. Hope became a projection toward future states. Loss became embedded in ritual mourning. The world was no longer only physical. It became psychological and metaphysical, expressed through symbols that carried layers of personal and collective meaning.

Yet this new capacity came with instability. Once the mind began to generate conceptual systems, those systems could conflict. Competing interpretations of the same symbol could fragment a group. Divergent narratives could reshape relationships. Misremembered events could generate entirely new traditions. Abstraction introduced cognitive power, and it introduced cognitive risk. The First Cognitive Revolution made human thought expansive, but it also made it fragile.

This fragility is central to understanding the deeper architecture of cognition. Collective Model Formation depends on stability. If individuals within a group interpret symbols differently or lose coherence in shared memory, the model weakens. This weakness did not matter in early societies with slow information flow. It matters profoundly today. The same architecture that once supported the emergence of culture now supports a global network of rapidly shifting ideas. The vulnerabilities introduced during the First Cognitive Revolution create openings that phenomena like Cognitive Drift can later exploit.

Drift emerges only in systems that rely heavily on internal representation. Early animals that lived through sensation alone could not drift. Humans, with symbolic worlds layered upon memory and imagination, created cognitive structures susceptible to disruption. The First Cognitive Revolution built the scaffolding of thought. It allowed the mind to climb into abstraction, but every layer added height and instability. The architecture grew taller, richer, and more delicate.

The First Cognitive Revolution was not a moment of sudden enlightenment. It was the beginning of a new cognitive ecology. Thought expanded into institutions, stories, rituals, and shared identities. The mind became capable of reorganizing itself through culture, a capability that shaped every advancement that followed. To understand human intelligence, we must understand this revolution. It marked the moment when thought stopped reflecting the world and began constructing worlds of its own.

PART - II
THE INVENTION OF MIND
HOW HUMANS BUILT INTERNAL WORLDS

4. Language: Humanity's First Technology

How Language Programs Thought

Human cognition changed irreversibly with the emergence of language. Before language, thought relied on sensory experience, emotional evaluation, and symbolic inference. These systems allowed early humans to imagine possibilities and recognize patterns, but they did not provide a stable framework for encoding abstract relationships or transmitting complex ideas. Language introduced that framework. It created a programmable structure for thought, a system that shaped how minds interpreted the world and how they organized internal experience. Cognitive scientists refer to this transformative mechanism as Linguistic Programming, the principle that language functions as the architecture through which thought is built, modified, and shared.

Linguistic Programming began when early humans discovered that sounds could be assigned to meanings that extended beyond immediate perception. A vocalization could refer to an object that was not present. A gesture could evoke a memory stored in the minds of others. A pattern of sounds could represent a prediction, a warning, or an instruction. This breakthrough separated thought from the constraints of the moment. It allowed minds to coordinate across time, distance, and experience. Language became the medium through which abstraction could move from an individual mind to the collective one.

The structure of language acted as a cognitive scaffold. It organized concepts into categories. It linked actions to consequences. It allowed sequences of events to be narrated, compared, and evaluated. Once these structures existed, they shaped the development of thought itself. Linguistic Programming reveals a crucial truth. Thought does not occur in isolation from language. Thought is molded by the linguistic forms available to it. A culture with words for specific emotions perceives those emotions more sharply. A society with a grammar that encodes time differently experiences time differently. Language does not simply label reality. It designs reality.

This design power created an immense cognitive advantage. Through language, early humans could compress complex ideas into shareable units. A single sentence could transmit the logic of a hunt. A story could encode moral expectations. A ritual chant could preserve generational knowledge. The mind gained tools for thinking that extended far beyond individual memory. These tools allowed humans to create models of the world that grew richer and more intricate with each generation.

Yet this power came with a new tension. Linguistic Programming does not only shape how individuals think. It shapes how groups think. Words create boundaries that can unify or divide. They define categories that become social truths. They create narratives that determine identity. Once a group agrees on a linguistic framework, that framework directs collective perception. This creates stability, but it also introduces rigidity. When language programs thought too strongly, interpretation becomes fixed. New insights struggle to emerge. The same system that expands cognition can begin to constrain it.

This tension is visible in the earliest linguistic traditions. Some groups used language to expand conceptual worlds. Others used it to enforce mythic or social hierarchies. Both uses relied on the same mechanism. Language structured cognition by determining what could be said and what could be imagined. Linguistic Programming shaped the limits of conceptual possibility.

The influence of language extended into the internal landscape of the mind. As children learned vocabulary, they learned categories. As they absorbed grammar, they absorbed structures for sequencing time, assigning agency, and interpreting causality. The mind became fluent in a code that determined how memories were organized and how relationships among ideas were constructed. Language became the template that guided internal simulation and reasoning. A thought could only exist in the form that language allowed it to take.

The relevance of Linguistic Programming stretches into the present era. Modern minds rely on linguistic models shaped not only by culture but by technology. Text, media, and communication platforms accelerate linguistic change. New categories appear. Old one's dissolve. Narrative structures that once stabilized cognition fragment into rapid, shifting forms. This transformation creates cognitive pressure. Linguistic Programming becomes dynamic and unstable. The mind must adapt to linguistic systems that evolve faster than it can internalize them.

This instability plays a central role in the emergence of Cognitive Drift. Drift affects systems that rely heavily on internal interpretation. Language is the primary vehicle for interpretation. When linguistic frameworks become destabilized, the thoughts built upon them become vulnerable to fragmentation. A shift in vocabulary can shift identity. A change in narrative context can alter memory. When language loses coherence, thought loses its anchor.

Language was humanity's first technology. It allowed minds to grow beyond the body and beyond the moment. It programmed cognition in ways that enabled culture, science, imagination, and cooperation. Yet its power carries a cost. The same structure that creates stable meaning can destabilize when overloaded or rapidly transformed. To understand modern cognition, we must understand this technology. Language programs the mind, and the mind lives inside the structures language creates.

Words as Memory, Imagination, and Power

The emergence of language did not only allow humans to communicate. It transformed the architecture of memory, expanded the reach of imagination, and created new forms of social and cognitive power. Words became more than sounds. They became instruments for shaping internal experience. This transformation is described by cognitive theorists as the Lexical Inheritance System, a framework in which words function as carriers of memory, imagination, and authority across individuals and generations. The Lexical Inheritance System explains how language reshaped the human mind by turning fleeting experiences into durable structures of thought.

Before words, memory existed as neural impressions tied to direct experience. A creature remembered danger because it felt danger. It remembered food because it consumed food. Memory was episodic and anchored to sensation. The arrival of words changed memory into something entirely different. A word could represent an event that occurred far away or long ago. It could preserve details that no individual had witnessed. It could transmit knowledge beyond the limits of personal experience. Words became containers that stored meaning at a resolution far greater than any single memory. The Lexical Inheritance System allowed information to accumulate across generations, creating a cognitive reservoir that shaped how individuals understood the world.

This shift had profound consequences for imagination. Once words stored not only events but possibilities, the mind gained a new domain in which imagination could operate. Words allowed early humans to construct hypothetical worlds, describe outcomes that had not occurred, or invoke entities that no one had seen. These imagined worlds were not mere fantasies. They served as cognitive laboratories. A hunter could imagine strategies before testing them. A community could envision seasons before preparing for them. Imagination became a structured exploration of potential realities, guided by the vocabulary available to the mind.

Yet imagination through language introduced a powerful tension. The worlds created by words could acquire an authority that rivaled physical reality. A story told often enough could become a memory. A belief repeated across generations could become a truth. Words could shape perception as forcefully as sensation. This influence created the conditions for both wisdom and illusion. The Lexical Inheritance System amplified cognitive reach, but it also opened the mind to forms of distortion that earlier species never encountered. A symbolic error could propagate across generations. A narrative bias could become an unquestioned rule. Words created stability, and they created fragility.

The power of language extended beyond internal cognition into social structure. Words allowed groups to define roles, establish norms, and coordinate collective behavior. Authority emerged from the ability to control meaning. A leader could shape narratives that unified a group or justified decisions. A ritual specialist could interpret symbols in ways that influenced collective emotion. This dynamic is known as Lexical Power Consolidation, the process by which control over words becomes control over minds. The individual who shapes language shapes the cognitive environment of the group.

Lexical Power Consolidation became a driving force in the early development of culture. Through shared vocabulary, communities created moral codes, origin myths, and systems of governance. These linguistic structures provided stability but also became weapons of influence. Words could elevate or silence. They could unify or divide. They could preserve knowledge or enforce ignorance. The mind, once guided by sensation and memory, became guided by vocabulary and narrative. The Lexical Inheritance System ensured that power in human societies was inseparable from control of language.

Modern cognition still operates within this system, but the scale has changed. Digital communication, mass media, and global networks amplify the reach of words beyond anything early humans could have imagined. Vocabulary shifts rapidly. Narratives rise and collapse within hours. Lexical Power

Consolidation now occurs across continents rather than villages. This acceleration creates instability within the cognitive environment. Minds struggle to adapt to linguistic systems that are continuously reshaping themselves. Interpretation becomes more demanding. Memory becomes more uncertain. The guidance once provided by stable vocabulary weakens.

This instability forms a crucial link to Cognitive Drift. Drift affects minds when interpretation becomes decoupled from coherent linguistic structure. If words lose clarity or meanings shift unpredictably, the thoughts built upon them become unstable. A narrative can fracture. A memory can distort. A belief can detach from its origins. The Lexical Inheritance System, once a stabilizing force in cognition, becomes a source of vulnerability when overwhelmed by rapid linguistic change.

Words shaped the human mind by allowing memory to expand, imagination to explore, and power to consolidate. They built the cognitive landscapes in which humans still live. Yet their influence carries both strength and risk. To understand how thought continues to evolve, we must understand the forces that shaped its earliest architecture. Words are not passive signs. They are active structures that shape what the mind remembers, imagines, and believes.

Story as the Engine of Human Reality

Human beings do not live only in the physical world. They live inside constructed realities shaped by stories. Stories organize time, assign meaning, and create coherence in experiences that would otherwise remain scattered. Through narrative, early humans learned to transform memory into guidance, imagination into strategy, and perception into shared understanding. This central role of narrative in cognition is known as Narrative Reality Construction, the process through which stories shape how the mind interprets the world. Narrative Reality Construction reveals that stories are not entertainment. They are cognitive engines.

In the earliest human communities, the world presented itself as a series of disconnected events. Weather changed without warning. Animals appeared and vanished. Life followed uncertain cycles. Without a unifying framework, these experiences remained random and overwhelming. Narrative provided the first tool for ordering this uncertainty. A sequence of events could be linked into cause and effect. A hunt could be remembered as a story instead of a list of sensory impressions. A death could become part of a myth rather than an isolated loss. Stories brought structure to a world that offered none.

This structuring power transformed memory. Through narrative, memory became selective and purposeful. Events were arranged into patterns that highlighted lessons, dangers, and opportunities. The mind learned to compress complexity into meaning. Narrative Reality Construction allowed memory to become a tool for prediction. A story about a past storm prepared the community for a future one. A story about a successful migration route encoded strategic knowledge. Memory gained shape through narrative, and through that shape it gained value.

Narrative also expanded imagination. Once humans developed the ability to tell stories that reached beyond lived experience, they could explore possibilities that no individual had encountered directly. A tale about distant lands encouraged exploration. A story about unseen forces encouraged ritual. A myth about ancestors encouraged continuity across generations. Stories created virtual worlds in which the mind could test ideas, emotions, and beliefs. This ability to explore hypothetical realities became one of the defining features of human cognition.

Yet narrative introduced a powerful tension. The same structure that created coherence could also generate distortion. A story that simplified reality could mislead. A narrative that emphasized danger could induce fear disproportionate to actual risk. A myth that strengthened group identity could divide groups from one another. Narrative Reality Construction enhances cognition, but it also shapes it selectively. What is excluded from a story becomes as powerful as what is included. Every narrative opens one path and closes others.

The social impact of narrative was equally profound. Stories became tools for coordinating group behavior. They defined norms, justified decisions, and preserved collective memory. Leadership often emerged from those who could tell the most compelling story rather than those with the greatest strength. Narrative persuasion allowed individuals to influence group direction. This influence formed the basis of Narrative Power Dynamics, the process by which control over stories becomes control over collective cognition. Narratives aligned communities, but they also established hierarchies based on who could shape the cultural imagination.

Narrative Power Dynamics expanded as societies grew larger. Oral traditions evolved into ritual performances, symbolic art, and eventually written myths. These systems reinforced shared frameworks that allowed thousands of individuals to coordinate as if they shared a single mind. Civilizations were built on stories that defined origins, purposes, and destinies. Narrative Reality Construction created cognitive scaffolding that supported the rise of culture, law, religion, and identity. To change the foundational stories of a society was to change the society itself.

The influence of narrative reaches deeply into the modern mind. Stories continue to structure memory, emotion, and decision making. They guide how individuals make sense of uncertain information. They determine which details are highlighted and which are dismissed. They stabilize identity by linking past, present, and future. Yet narrative today operates within an environment of unprecedented acceleration. Digital platforms reward short, emotionally charged stories. Narratives compete and mutate at a pace that overwhelms cognitive stability. Narrative Power Dynamics have expanded into global networks, where millions of minds are shaped by stories circulating at high speed.

This acceleration creates a pathway for Cognitive Drift. Drift thrives when narrative coherence weakens. A disrupted narrative can alter memory, shift identity, or fracture collective understanding. When stories lose structure, the world becomes harder to interpret. The mind struggles to maintain continuity.

Narrative Reality Construction, once a stabilizing force, becomes vulnerable to distortion.

To understand human cognition, we must understand the central role of story. Stories do not describe reality. They produce it. They build the mental frameworks through which the world becomes meaningful. They create the engines that power thought.

5. Consciousness and the Birth of the Self

What Awareness Solves

Consciousness did not emerge to contemplate beauty or to unravel the mysteries of existence. It emerged to solve problems that instinct, memory, and perception alone could not handle. Early organisms reacted to the world. Early mammals interpreted the world. Early humans began to imagine the world. Yet imagination introduced complexity. A mind that could envision multiple futures also needed a mechanism to evaluate them, prioritize them, and choose among them. This requirement produced one of evolution's most astonishing innovations, a process cognitive scientists describe as the Awareness Integration System. The Awareness Integration System is the set of mechanisms through which consciousness synthesizes perception, memory, emotion, and intention into a coherent field that can guide action.

Before awareness, cognition was fragmented. Signals arrived through different sensory channels. Memories existed as traces scattered across neural circuits. Emotional responses triggered behavioral shifts without any central coordination. The result was an organism that reacted effectively but lacked a unified representation of its own state. The emergence of awareness changed this fragmentation. Awareness created a workspace where information from multiple sources could be integrated, compared, and transformed. It allowed the organism to know not only what it perceived but how it felt about what it perceived. Awareness provided context.

This context was essential for navigating uncertainty. A creature capable of awareness could evaluate internal states in relation to external threats. Hunger could be weighed against caution. Curiosity could be balanced against risk. Awareness allowed an organism to monitor its own priorities and adjust them moment by moment. This ability solved a fundamental evolutionary problem. It produced flexible decision making that was no longer bound to automatic responses. The Awareness Integration System gave cognition the capacity to pause, to reconsider, and to select.

Awareness also addressed the challenge introduced by Prospective Imagination. Once minds began generating internal futures, they needed a mechanism to determine which imagined paths were valuable and which were dangerous. Awareness created this evaluative layer. It monitored the internal landscape and distinguished between simulations that aligned with reality and those that could lead to harm. Awareness helped prevent the mind from becoming trapped in its own imagined possibilities. It grounded imagination within adaptive limits.

This grounding introduced a new form of cognitive tension. Awareness increased the power of the mind, but it also increased its vulnerability. A system that monitors itself can become overwhelmed by its own complexity. Awareness made it possible for cognition to recognize conflict, doubt, and contradiction. These states had no meaning in organisms without awareness. In humans, they became central experiences. The Awareness Integration System brought coherence, but coherence required continuous effort.

As awareness refined, it expanded beyond immediate survival. Humans began to recognize not only their internal states but their existence as enduring beings. Awareness stretched across time. It made individuals capable of linking past experiences to future intentions. This temporal integration created continuity. A person who remembered their childhood and imagined their old age occupied a psychological timeline that earlier minds could not access. Awareness solved the problem of identity across time. It provided the sense of being the same entity from one moment to the next.

This continuity made complex cooperation possible. Group members who recognized each other as stable individuals could form long term bonds. They could create shared plans and maintain trust. Awareness allowed humans to attribute intentions and feelings to others, enabling empathy and coordination. The Awareness Integration System extended from the self to the social world, creating a shared space of understanding that transformed human interaction.

Yet this architecture introduced a new vulnerability. Awareness made the mind dependent on internal coherence. When coherence is disrupted, cognition destabilizes. Conflicting memories, fractured narratives, or shifting emotional states can undermine the sense of continuity that awareness provides. This vulnerability is a precursor to phenomena like Cognitive Drift. Drift affects systems that rely on integrated awareness. When awareness fails to align perception, memory, and interpretation, the internal field begins to distort. Drift exploits the gaps where coherence weakens.

Awareness solved profound evolutionary challenges. It unified fragmented cognitive processes, guided decision making, regulated imagination, and created continuity of self. It made empathy possible and enabled the complex social structures that define human life. Yet its strengths reveal its fragility. The Awareness Integration System remains essential for navigating the world, but it is also susceptible to overload and disruption. Understanding what awareness solves allows us to understand what threatens it.

Inner Narrative and Identity Formation

Awareness created coherence, but coherence alone could not sustain a stable sense of self. As human cognition expanded, the mind needed a structure that could organize memories, emotions, and intentions into a unified psychological identity. This structure emerged as the inner narrative, a continuous stream of interpretation through which individuals explain their experiences and define their place in the world. Researchers refer to this process as Narrative Self-Assembly, the mechanism by which consciousness constructs an internal story that becomes the foundation of identity.

Narrative Self-Assembly begins with the simple act of linking events. A child remembers falling, then remembers rising, and connects those moments into a sequence that implies learning. This sequence becomes a story, and the story becomes part of how the child perceives themselves. The mind discovers that events gain meaning only when arranged into a narrative arc. This arc provides direction and purpose. It allows the self to imagine a future that extends from the past. Without narrative, consciousness becomes a collection of moments without continuity.

The inner narrative does more than preserve memory. It interprets memory. When the mind retrieves a past experience, it rarely recalls the raw event. Instead, it retrieves the story that was created around it. This story reflects not only what happened but what the individual believes the event meant. Narrative Self-Assembly gives memory a shape that aligns with the person's evolving identity. This shaping allows the mind to maintain internal consistency, but it also introduces selective interpretation. Some details are highlighted. Others are discarded. The story stabilizes identity, but stability comes at the cost of accuracy.

As narrative complexity grows, it becomes the lens through which individuals interpret the world. Every new experience is compared to the story already in progress. A rejection becomes part of a story about inadequacy, empowerment, or resilience. A success becomes part of a story about capability or transformation. The narrative reaches forward and backward, influencing how individuals imagine the future and remember the past. Narrative Self-Assembly turns consciousness into an unfolding novel in which the mind is both author and protagonist.

This narrative structure plays a crucial role in social life. People present their stories to others through conversation, expression, and behavior. These external narratives influence how groups perceive individuals and how individuals perceive themselves. Social identity becomes intertwined with personal narrative. The story a person tells is rarely entirely their own. It is shaped by culture, expectations, and interactions with others. Identity becomes a negotiation between internal narrative and external recognition. The self is not a fixed entity. It is a story told across minds.

This dynamic introduces a significant cognitive tension. Narrative Self-Assembly requires coherence, yet life rarely provides coherence. Contradictions, ambiguities, and unexpected events challenge the stability of the internal story. When the narrative cannot adapt to new experience, psychological instability emerges. Individuals may reinterpret memories to preserve the narrative or ignore evidence that threatens their sense of self. This defensive rewriting protects identity, but it can distort cognition. The narrative becomes rigid, and the mind becomes fragile.

The fragility of narrative identity is especially visible in moments of disruption. A sudden loss can fracture the story. A major failure can rewrite it. A profound achievement can shift its direction entirely. These disruptions expose the underlying mechanism. Narrative Self-Assembly is not passive reflection. It is active construction. The mind works continuously to maintain the illusion of a stable self. This illusion is functional, but it is also precarious.

This precarity has direct relevance to Cognitive Drift. Drift emerges when the systems that maintain narrative coherence begin to destabilize. If memories lose their context or interpretations shift unpredictably, the inner narrative becomes unreliable. A person may sense that they are the same individual yet feel that parts of their story no longer fit. The continuity that once grounded identity begins to loosen. Drift exploits the instability created by narrative tension.

Identity formation begins with awareness, but it matures through narrative. The inner story gives shape to memory, meaning to experience, and direction to life. It allows the mind to navigate uncertainty by providing a framework for interpretation. Yet it is also a delicate structure, held together by selective recall and imaginative reconstruction. The self is a story the mind tells to create order, and that story is always one disruption away from transformation.

The Useful Illusion of "Me"

Human consciousness produces a powerful impression. It feels as if there is a single enduring entity at the center of experience, a unified self that thinks, chooses, remembers, and acts. This sense of continuity is compelling, but it is not a literal description of how the mind works. Cognitive science reveals that the self is not a fixed substance. It is a dynamic construction generated by multiple interacting systems. Researchers refer to this construction as the Identity Illusion Mechanism, the process through which the mind produces the feeling of a stable "me" even though no such entity exists in any single location. The self is not an object. It is an interpretation.

The Identity Illusion Mechanism emerges from the need for coherence within a complex cognitive system. Perception, memory, emotion, and decision making operate across distributed neural circuits. Each circuit contributes information that must be coordinated if the organism is to act effectively. Awareness provides a unified field for these signals, but awareness alone does not create identity. Identity requires a narrative framework that interprets these signals as belonging to the same agent across time. Narrative Self-Assembly supplies this framework. Together, these processes create an illusion that is extraordinarily useful. The mind believes there is one continuous self because such belief stabilizes behavior.

This illusion solves several evolutionary problems. A stable sense of self allows long term planning. It supports moral learning and social accountability. It enables empathy by allowing individuals to imagine the experiences of others through analogy with their own. The Identity Illusion Mechanism makes cooperation more predictable and decision making more consistent. Without this mechanism, the mind would behave like a collection of conflicting impulses rather than a coordinated agent. The illusion is not a flaw. It is a feature designed to transform fragmentation into unity.

Yet this unity is constructed moment by moment. The mind continually updates its model of the self-based on memory retrieval, emotional context, and social feedback. A single event can shift self-interpretation. A failure can weaken confidence. A success can alter expectations. These shifts reveal the fluid nature of identity beneath the illusion of stability. The self is not a completed structure. It is a process of ongoing revision. The Identity Illusion Mechanism conceals this constant reconstruction, allowing experience to feel continuous even when the underlying architecture is constantly changing.

This concealment introduces cognitive tension. The mind depends on the illusion for stability, yet the illusion is fragile. When memories conflict or emotional states shift rapidly, the sense of self can feel disrupted. Individuals may question who they are or what they believe. These disruptions expose the machinery behind the illusion. They reveal that identity is not discovered but continuously manufactured. The same processes that create the self can undermine it when coherence weakens.

Social environments amplify this fragility. Other minds influence the stories individuals tell about themselves. Praise, rejection, belonging, and exclusion all shape the internal narrative. Culture provides templates for identity, and individuals adapt their stories to fit these templates. When cultural contexts shift, the narrative must shift with them. The self is not autonomous. It is relational. Identity is shaped as much by external interpretation as by internal construction. The Identity Illusion Mechanism depends on both.

This dependence has deep implications for cognitive vulnerability. When the systems that maintain the illusion begin to destabilize, the sense of self becomes uncertain. Contradictory memories, fragmented narratives, or shifting meanings can weaken the illusion of continuity. This weakening provides an entry point for phenomena like Cognitive Drift. Drift exploits instability in the internal model of the self. When the narrative no longer aligns with experience, identity feels fractured. The illusion that once unified cognition becomes a site of disruption.

Understanding the illusion of "me" does not diminish its value. The illusion is essential. It provides the structure through which humans navigate the world. It gives meaning to choices and continuity to life. Yet recognizing its constructed nature reveals the delicate balance at the core of consciousness. The Identity Illusion Mechanism allows the mind to function as a coherent agent, but it does so by simplifying a complex, distributed system. The self is not a singular truth. It is a useful illusion, crafted by the mind to transform chaos into order.

6. Tools, Fire, and Externalized Thought

Technology as Cognitive Amplifier

Human cognition did not evolve in isolation. It grew through interaction with the physical world and through innovations that transformed the limits of thought. Early humans discovered that tools were not merely objects. They were extensions of the mind. A sharpened stone increased the power of the hand. A woven basket increased the reach of memory by storing future resources. Fire expanded the boundaries of night and reshaped the rhythms of life. These discoveries produced one of the most consequential developments in cognitive evolution, a process known as the External Amplification Principle. The External Amplification Principle describes how technology magnifies cognitive capacity by shifting mental operations into the environment.

Before the emergence of tools, cognition was bound to the constraints of the body. Tasks required physical effort, memory relied on biological storage, and planning depended entirely on internal simulation. The arrival of tools disrupted these limits. A tool preserved effort. Fire preserved warmth. Containers preserved abundance. Each innovation changed what the mind needed to compute. The world became a partner in cognition rather than a passive setting. Technology began to handle part of the cognitive load.

This shift produced a feedback cycle. A tool reshaped behavior. New behaviors created new cognitive demands. These demands inspired further innovation. The External Amplification Principle turned the environment into a second cognitive system. A spear extended reach. A net extended coordination. A calendar etched on bone extended temporal awareness. The mind no longer operated alone. It interacted with structures that encoded memory, strategy, and prediction outside the skull.

One of the earliest demonstrations of cognitive amplification occurred through the control of fire. Fire reshaped diet, expanded communal activity, and altered sleep cycles. Cooked food reduced digestion time and increased caloric availability, which supported brain growth. Extended light allowed longer periods for communication, ritual, and planning. Fire also fostered social cohesion. People gathered around it and shared stories, creating early forms of collective memory. A simple technology became a force that reorganized cognition at individual and group levels.

As tools multiplied, their influence deepened. The earliest stone blades required planning to create. They required attention to geometry, force, and fracture patterns. Toolmaking trained the brain to anticipate consequences, refine precision, and evaluate feedback. This training influenced neural development. It shaped hand control, visual reasoning, and the integration of multi-step processes. Through the External Amplification Principle, tools sculpted the very circuits that later supported language and symbolic thought.

The evolution of external storage systems intensified this relationship. Marks on bone, painted symbols, and carved notches allowed information to persist beyond a single mind. These systems were not yet writing, but they served as cognitive scaffolding. They extended memory into physical space. The mind relied on these external traces to recall events, organize tasks, and coordinate groups. External storage reduced the burden of biological memory and freed cognitive resources for abstraction and innovation.

Yet technology did more than amplify cognition. It introduced tension into cognitive evolution. Each innovation increased human capability, but it also increased dependence on external structures. As tools grew more complex, the mind became increasingly shaped by the systems it created. The External Amplification Principle gave humans extraordinary power, but it also reduced the autonomy of cognition. The mind began to rely on environmental extensions that it could not function without. This interdependence strengthened cognition and weakened it simultaneously.

This tension foreshadows modern cognitive instability. When a mind relies on external amplifiers, disruptions in those amplifiers can destabilize cognition. A break in communication tools weakens coordination. A loss of stored information disrupts planning. A shift in technological systems alters attention and memory. The External Amplification Principle explains why modern thought, heavily augmented by digital systems, can become vulnerable to overload and fragmentation. Technology expands cognition, and it exposes cognition.

Cognitive Drift arises in systems where internal processes and external supports become misaligned. In early human evolution, tools and fire created stable forms of amplification. In the modern world, the pace of technological change makes alignment difficult. Minds struggle to adapt to systems that evolve faster than biological cognition can recalibrate.

Technology has always been a cognitive amplifier. It strengthened the hand, the memory, the imagination, and the social bond. Yet amplification is never neutral. It shapes the architecture of thought as much as though shapes it. To understand the future of intelligence, we must understand the principle that has guided human evolution for hundreds of thousands of years. Our minds have always grown in partnership with our inventions.

The Mind–Tool Feedback Loop

Human evolution did not unfold through biological change alone. It advanced through an ongoing exchange between cognition and the tools that cognition produced. Each innovation altered how the mind perceived the world, solved problems, and organized experience. These cognitive changes then generated new forms of technology that continued the cycle. This dynamic interaction forms what researchers describe as the Cognitive–Technological Feedback Loop, a self-reinforcing process in which tools shape the mind and the mind shapes tools. The Cognitive–Technological Feedback Loop is one of the central engines of human advancement and one of the origins of modern cognitive instability.

The earliest tools demonstrate the power of this loop. When early humans shaped stone into blades, the process required precision, planning, and anticipation. These cognitive demands changed the brain itself. Neural circuits responsible for fine motor control and spatial reasoning strengthened. The improved cognitive capacity then enabled the creation of more sophisticated tools. Each advancement in technology reflected a corresponding advancement in cognition. The Cognitive–Technological Feedback Loop gradually produced a mind capable of far greater abstraction and coordination than any other species.

This loop intensified with the emergence of composite tools. Spears, hafted axes, and woven nets required the integration of multiple materials, each with distinct properties and constraints. The design of these tools forced early humans to think in terms of systems rather than isolated objects. They learned to anticipate how components interacted. They learned to predict how the tool would behave under different conditions. This predictive capacity deepened cognition, encouraging the development of mental models that extended beyond immediate perception. Cognitive architecture evolved alongside material innovation.

The loop grew even more powerful with the emergence of external memory systems. Marks, notches, and symbolic patterns allowed information to persist outside the human brain. Once information could be stored externally, cognition no longer needed to rely solely on biological memory. This transformation freed cognitive resources for tasks such as planning, storytelling, and conceptual exploration. External memory systems expanded the reach of the mind, but they also increased the influence of tools on thought. The Cognitive–Technological Feedback Loop became a partnership in which tools not only extended cognition but actively shaped the architecture of reasoning.

As the loop continued, its influence spread beyond technical skills into the deeper layers of culture and identity. Tools changed social structures by enabling new forms of cooperation. Hunting strategies grew more complex. Division of labor expanded.

Shared rituals emerged around technological practices such as fire craft or toolmaking. These social changes further reshaped cognition. The mind evolved in environments structured by the very tools it had created. Each new tool introduced new possibilities and new constraints. The Cognitive–Technological Feedback Loop produced not only external innovation but internal transformation.

This interdependence introduced a fundamental tension. As cognition became increasingly reliant on tools, the mind gained power but lost autonomy. A tool that failed at a crucial moment endangered survival. A tool that demanded complex coordination required stable social structure. A tool that preserved memory could distort interpretation. The Cognitive–Technological Feedback Loop created a system in which cognitive strength and cognitive vulnerability increased together. Tools allowed humans to exceed biological limitations, but they also exposed the mind to new kinds of fragility.

In the modern era, this tension has intensified dramatically. Digital systems amplify attention, memory, and decision making in ways that no previous technology achieved. These systems operate at speeds far beyond biological processing. They present information in volumes that exceed natural cognitive capacity. The Cognitive–Technological Feedback Loop now accelerates at a pace that challenges the mind's ability to adapt. External systems influence not only how humans think but what they think. Technology no longer waits for cognition to adjust. It evolves faster than the biological substrate can follow.

This acceleration creates fertile ground for Cognitive Drift. Drift emerges when internal and external cognitive systems fall out of alignment. In earlier eras, tools evolved slowly, allowing biological cognition to recalibrate. Today, the gap widens with every technological shift. Memory offloaded to devices weakens internal recall. Digital narratives reshape identity. Algorithmic systems influence attention patterns. The Cognitive–Technological Feedback Loop that once drove human progress now creates pressure points that destabilize thought.

Understanding this loop is essential for understanding the future of cognition. Human intelligence is not a closed biological system. It is a hybrid architecture shaped by the interplay of brain, body, and tool. The Cognitive–Technological Feedback Loop reveals why humans continue to innovate and why innovation continues to reshape the mind. It explains both the strengths that carried the species forward and the vulnerabilities that now define its cognitive environment. Tools changed the mind. The mind changed tools. Together, they created the world we inhabit.

Intelligence Beyond the Skull

Human intelligence did not stop at the boundaries of the brain. As cognition expanded through language, tools, and symbolic systems, the mind learned to distribute its functions into the environment. This distribution was not accidental. It was an evolutionary strategy that allowed humans to overcome the biological limitations of memory, attention, and reasoning. Cognitive theorists describe this transformation as the Extended Intelligence Architecture, the system through which thought operates across brain, body, and external structures. The Extended Intelligence Architecture reveals that much of what we call human intelligence exists outside the skull.

The earliest forms of extended intelligence appeared through simple technologies that stored information in physical space. A carved line marked a season. A pile of stones marked a path. A painted symbol marked a memory. These representations were not merely reminders. They were components of cognition. They carried knowledge that no individual brain needed to retain. They shaped perception and guided action. Through these external traces, early humans offloaded cognitive tasks into the environment, freeing mental resources for imagination and strategic thinking.

The Extended Intelligence Architecture grew more sophisticated as symbolic systems expanded. Tokens represented quantities. Marks represented narratives. Ritual objects represented collective meaning. These symbolic tools functioned as cognitive scaffolding that supported complex reasoning. They allowed individuals to think about abstractions without holding every detail in working memory. They allowed groups to synchronize their understanding of events, identities, and obligations. The world became structured with cues and symbols that guided human thought as reliably as neural circuits.

This integration deepened with the emergence of writing. Writing created a durable cognitive substrate that preserved ideas across time and space. The mind developed the ability to interact with its own recorded outputs. A written calculation could refine future calculations. A written story could reshape cultural imagination. A written law could influence behavior long after its author had disappeared. Writing created cognitive continuity at a scale that biology alone could never provide. The Extended Intelligence Architecture became intergenerational.

As societies grew, so did their external cognitive systems. Maps expanded spatial intelligence. Calendars expanded temporal intelligence. Architecture encoded cosmology and social hierarchy. Institutions encoded norms, knowledge, and expectations. These structures served as extensions of human reasoning, storing information that guided perception and behavior. The mind no longer needed to compute everything internally. It relied on environments shaped by human intention. Intelligence expanded outward into the physical and social world, becoming a distributed process rather than a solitary one.

Yet this expansion introduced a critical tension. The more intelligence moved into external structures, the more cognition depended on the stability of those structures. A collapse of records could erase knowledge. A disruption in symbolic systems could fracture collective identity. A shift in cultural narratives could alter the interpretive frameworks that guided thought. The Extended Intelligence Architecture strengthened cognition, but it created vulnerabilities that were not present in earlier minds. External intelligence required external stability.

This tension escalated with the arrival of digital systems. Digital environments became cognitive partners that handled memory, navigation, prediction, and communication. These systems processed information at scales that exceeded biological capacity. They began to influence attention patterns, emotional responses, and decision making. The Extended Intelligence Architecture entered a new phase in which external systems were no longer passive storage devices. They became active participants in cognition. Intelligence beyond the skull became dynamic, adaptive, and algorithmically driven.

This dynamism introduced instability. Digital systems evolve faster than biological cognition can adapt. They restructure information flow, reshape narrative context, and accelerate the pace of interpretation. Minds that rely on these systems face constant shifts in cognitive landscape. The Extended Intelligence Architecture moves from stability to flux. This flux creates fertile conditions for Cognitive Drift. Drift emerges when the alignment between internal cognition and external structures breaks. A disrupted ecosystem of tools, symbols, or narratives destabilizes thought. Interpretation becomes uncertain. Identity becomes fluid. Continuity weakens.

Understanding intelligence beyond the skull is essential for understanding the modern mind. Human cognition is not a sealed chamber of neurons. It is a distributed architecture that spans tools, symbols, environments, and networks. This architecture enabled the rise of culture, science, and civilization. It allowed humans to think in ways no biological brain could achieve alone. Yet it also introduced dependencies that now shape the vulnerabilities of thought. The Extended Intelligence Architecture is both the source of human power and the origin of human fragility.

PART - III

CULTURE: THE SECOND EVOLUTION OF THOUGHT

HOW SOCIETIES BECAME COGNITIVE SYSTEMS

7. How Culture Thinks for Us

Shared Myths as Cognitive Infrastructure

Human societies did not grow through numbers alone. They grew through shared meaning. Early humans faced a world filled with uncertainty, danger, and limited resources. To survive at scale, they needed more than tools and language. They needed a way to coordinate expectations, unify behavior, and create trust among individuals who did not share family ties. This requirement produced one of the most influential developments in cognitive evolution, a phenomenon known as Mythic Framework Construction. Mythic Framework Construction describes how shared myths became cognitive infrastructure, shaping perception, guiding behavior, and binding individuals into coherent groups.

Shared myths were not fictional distractions. They were functional systems. A myth created a common explanatory model for the world, offering reasons for natural events, moral rules for behavior, and narratives that linked individuals to a larger purpose. These stories provided continuity across generations, allowing societies to transmit values and knowledge in ways that were emotionally resonant and cognitively efficient. Myths turned scattered experiences into structured understanding.

The power of Mythic Framework Construction lay in its ability to synchronize minds. Individuals who believed in the same origins, destinies, and moral codes could cooperate without constant negotiation. The myth became a shared mental model that shaped interpretation automatically. A ritual symbol meant the same thing to everyone present. A sacred object carried a shared emotional weight. A cosmological story defined the boundaries of right and wrong. These shared models reduced cognitive load. People did not need to calculate social expectations. The myth carried the calculation for them.

This synchronizing effect created a profound evolutionary advantage. Groups with strong shared myths could coordinate at scales far beyond kinship. They could build structures, defend territory, and organize labor. They could imagine futures that extended beyond individual lifespans. Mythic Framework Construction allowed societies to align internal narratives with collective objectives. The story that governed the group governed the individual mind.

Yet shared myths introduced a subtle tension. The more powerful they became as cognitive infrastructure, the more they shaped perception. Myths did not describe reality. They defined it. They filtered sensory information, simplified complexity, and assigned significance where the world itself was ambiguous. This filtering allowed individuals to navigate uncertainty, but it also limited alternative interpretations. A myth that united could also constrain. A narrative that guided could also blind.

This tension grew as societies expanded. Different groups developed different mythic frameworks, each shaping cognition in distinct ways. Conflicts between groups were often conflicts between incompatible cognitive infrastructures. When two societies believed in different origins or destinies, cooperation became fragile. Mythic Framework Construction strengthened internal cohesion but intensified external division. The same mechanism that built trust within a group created distrust between groups.

Myths also influenced emotional life. By embedding meaning in symbols and narratives, they shaped how individuals experienced fear, hope, obligation, and identity. A myth could elevate courage or amplify guilt. It could bind individuals to ancestors or to future generations. These emotional structures made myths durable, but they also made them resistant to revision. When a myth became intertwined with identity, challenging the myth felt like challenging the self.

The cognitive power of myths is still visible today. Modern ideologies, nations, and institutions rely on mythic frameworks disguised as historical narratives or cultural truths. These frameworks guide interpretation, shape collective behavior, and define the boundaries of belonging. They act as cognitive infrastructure even when their origins are forgotten. Mythic Framework Construction persists because it solves the same problem it solved for early societies. It provides shared meaning in a complex world.

But this infrastructure introduces vulnerability in the modern era. When myths no longer align with reality, the cognitive guidance they offer becomes unstable. Narratives fracture. Consensus weakens. Social trust dissolves. Minds that rely on shared myths to interpret the world begin to drift into incompatible realities. This fragmentation creates fertile ground for Cognitive Drift. Drift emerges not only within individuals but within societies when shared cognitive infrastructure breaks down.

Understanding shared myths as cognitive infrastructure reveals the profound influence of narrative on human evolution. Myths build societies by building minds. They shape perception, guide emotion, and coordinate behavior. They solve the problem of collective meaning, yet they carry the risk of collective distortion. To understand culture, we must understand the frameworks that think for us.

Ritual as a Social Memory System

Long before humans created archives, written records, or formal institutions, they preserved knowledge through repeated action. Rituals were not decorative traditions or symbolic performances. They were engineered patterns of behavior that stored information in the bodies of individuals and in the rhythms of collective life. Anthropologists refer to this process as Ritual Memory Encoding, the mechanism through which societies embed essential knowledge into coordinated, repeated practices that transmit meaning across generations without written language.

Ritual Memory Encoding operates by converting abstract cultural principles into sensory and motor routines. Through repetition, these routines become deeply embedded in neural circuits, allowing communities to retain complex information even when cognitive resources are limited. A ritual could encode seasonal cycles, moral rules, social hierarchies, or cosmological beliefs. By performing a ritual, individuals learned not only the action but the meaning behind the action. The ritual became a living memory system that preserved the identity of the group.

For early societies, this encoding solved a fundamental cognitive problem. Without external storage devices, information could be lost within a generation. A drought, a migration, or a conflict could break the chain of oral transmission. Rituals created redundancy by distributing memory across many bodies. When knowledge was enacted rather than spoken, its survival did not depend on a single storyteller. It could be preserved through collective participation.

The power of ritual lay in its multisensory nature. Movement, sound, rhythm, color, and touch worked together to create a vivid cognitive imprint. A chant reinforced communal identity. A dance encoded social roles. A shared meal transmitted values of unity or hierarchy. Rituals used emotional intensity to strengthen memory. Joy, fear, reverence, and solemnity heightened attention, ensuring the ritual would not be forgotten. Through these mechanisms, Ritual Memory Encoding produced learning that was automatic, durable, and difficult to erase.

As societies grew larger, rituals became essential for maintaining cohesion. They synchronized attention and emotion, aligning individuals with the group's internal logic. When people gathered for a ritual, they felt the presence of the collective mind. The ritual created a shared psychological state, collapsing individual differences and emphasizing group identity. These states enhanced trust, reduced internal conflict, and coordinated expectations. The ritual did not simply transmit memory. It transmitted belonging.

Yet this mechanism introduced cognitive tension. Rituals preserved stability, but stability can become rigidity. As environmental or social conditions changed, rituals often remained the same. The memory encoded in ritual was preserved so strongly that it resisted adaptation. A ritual that once protected the group could become a barrier to progress. The same emotional force that strengthened memory could prevent innovation. Ritual Memory Encoding ensured continuity, but continuity sometimes conflicted with relevance.

This tension is especially visible when rituals become detached from the original information they encoded. Individuals may continue performing actions long after the meaning is forgotten. The ritual becomes a ghost memory, a behavior that persists without context. These ghost memories can shape identity and behavior in ways that are no longer aligned with the group's needs. The ritual still influences cognition, but its purpose is obscured.

Modern societies retain this dynamic. National ceremonies, religious practices, professional traditions, and even digital habits function as ritualized behaviors that encode shared values and expectations. Graduation ceremonies encode achievement. Courtroom procedures encode authority. Online trends encode belonging. These rituals guide interpretation and reinforce identity, even when participants do not consciously reflect on their meaning.

However, the modern environment accelerates the pace of change. Rituals that once aligned with social realities struggle to keep up with shifting cultural narratives. When rituals lose relevance, the cognitive guidance they provide weakens. People begin to interpret the same ritual in divergent ways or abandon it entirely. As collective memory fragments, shared meaning dissolves. This dissolution creates openings for Cognitive Drift. Drift emerges when the systems that once synchronized cognition no longer function reliably. Without stable ritual structures, societies lose the rhythms that kept their interpretations aligned.

Ritual as a social memory system reveals how deeply culture shapes cognitive evolution. Rituals preserve information not by storing it on surfaces but by embedding it in collective behavior. They create continuity, identity, and shared understanding. Yet they are also vulnerable to misalignment and decay. To understand how societies, think, we must understand how rituals store their memories.

Society as a Distributed Mind

Human intelligence does not reside solely within individuals. It emerges from the interactions, agreements, conflicts, and coordinated actions of many minds working together. When humans formed groups, they created systems capable of solving problems that no single brain could solve alone. Anthropologists and cognitive theorists refer to this phenomenon as Distributed Societal Cognition, the process by which societies function as collective minds that store knowledge, generate insight, and regulate behavior across interconnected individuals.

Distributed Societal Cognition begins with the simple act of coordination. When individuals cooperate, they share information, divide labor, and synchronize attention. These interactions produce cognitive structures larger than any single mind. A hunting party does not merely combine bodies. It combines perception, strategy, and memory into a unified decision making system. A community gathering does not simply strengthen social ties. It aligns emotional states, moral expectations, and collective understanding. Through cooperation, individual minds merge into a dynamic, distributed intelligence.

This merging becomes more powerful as societies grow. Knowledge accumulates across generations, preserved in tools, rituals, myths, and institutions. No individual needs to understand the entire system. Each person carries a fragment of the collective mind. A craftsperson knows techniques that were refined by centuries of trial and error. A healer knows remedies inherited from long forgotten ancestors. A leader interprets norms shaped by countless historical events. Distributed Societal Cognition allows a society to solve complex problems without any single mind holding the full blueprint.

Shared institutions amplify this effect. Courts, markets, councils, and religious systems function as cognitive modules that stabilize interpretation and regulate behavior. These institutions store memory, enforce norms, and guide collective decision making. Individuals rely on these structures to navigate uncertainties that exceed personal knowledge. The society becomes a cognitive ecosystem in which information flows across roles and generations. The distributed mind grows with every contribution, every correction, and every shared experience.

However, Distributed Societal Cognition introduces a deep cognitive tension. The collective mind must remain coherent enough to coordinate behavior, yet flexible enough to adapt to new realities. Too much rigidity leads to collapse under changing conditions. Too much flexibility leads to fragmentation. The balance between unity and adaptability becomes a defining challenge for societies. When the collective mind holds together, societies flourish. When it fractures, societies enter conflict, confusion, and decline.

This tension becomes sharper when groups develop competing distributed minds. Each society constructs its own cognitive ecosystem through myths, rituals, laws, and institutions. These systems can be incompatible with one another. A moral principle that stabilizes one society can destabilize another. A narrative that unifies one group can provoke resistance in a different group. Distributed Societal Cognition allows societies to think, but it also causes societies to think differently. These divergent cognitive systems generate conflict not merely over resources but over meaning.

The evolution of writing expanded distributed cognition even further. Texts allowed societies to store vast amounts of information outside biological memory. Laws became durable. Histories became reference points. Scientific knowledge accumulated across centuries. Writing allowed the collective mind to refine itself, correct errors, and build upon previous insights. The societal mind became cumulative and self-improving.

Modern technology pushes this phenomenon to unprecedented levels. Digital networks enable instant sharing of information across continents. Social platforms synchronize emotional and cognitive states at massive scales. Collective interpretation now happens in real time. Distributed Societal Cognition becomes faster, broader, and more volatile. The shared mind of a nation or a global community can form, dissolve, or fracture within hours. These accelerated dynamics reshape human thought at both individual and societal levels.

This acceleration introduces new vulnerabilities. When the distributed mind becomes unstable, individuals lose the cognitive support that society once provided. Institutions can no longer guarantee shared meaning. Myths and rituals fragment. Collective memory becomes inconsistent. As coherence weakens, groups drift into separate interpretive worlds. This fragmentation creates fertile conditions for Cognitive Drift. Drift emerges not only within individuals but at the societal level when the distributed mind splinters into incompatible realities.

Understanding society as a distributed mind reveals the profound interdependence between individual cognition and collective structures. Human intelligence is not a solitary achievement. It is the outcome of millions of minds linked through shared practices, narratives, technologies, and institutions. The distributed mind gives rise to culture, science, and civilization, yet it remains vulnerable to fragmentation and instability. To understand the future of human thought, we must understand the collective intelligence that has always shaped it.

8. Writing and the Birth of Recorded Mind

When Memory Moved Outside the Brain

Human cognition reached a turning point when memory no longer lived solely within biological tissue. For hundreds of thousands of years, the brain carried the full burden of remembering. Stories, rituals, and social structures preserved fragments of the past, but these systems depended on people, and people forgot. Knowledge could vanish in a single generation. The emergence of writing changed this trajectory completely. Writing created a new form of cognition, a phenomenon scholars call External Memory Transfer. External Memory Transfer is the process through which human memory relocates into symbols, objects, and surfaces that exist independently of the mind that created them.

The transformation began with marks that were not yet language but served as memory anchors. A series of lines carved into bone recorded lunar cycles. A symbol painted on stone preserved a hunting path. A pattern etched into pottery captured the identity of a household. These early inscriptions were not storytelling devices. They were tools for stabilizing memory in a world where forgetting carried real consequences. Once information could be placed outside the mind, it no longer depended on fragile neural networks. The environment became a partner in cognition.

This transition solved a profound evolutionary problem. Biological memory is limited. It decays, distorts, and competes with new information for space. External Memory Transfer created a storage system that did not decay with age or vanish with death. A symbol carved into stone endured beyond the individual. Knowledge became transferable across time on a scale impossible for oral societies. A community no longer needed to rely on the most skilled storytellers to preserve essential information. The world itself held the record.

As writing evolved into structured systems, its power intensified. Symbols became standardized. Sequences became meaningful. Scribes learned to encode transactions, laws, genealogies, and cosmologies. Writing created a stable cognitive backbone that allowed societies to scale beyond the limits of biological memory. Complex administration became possible. Stored knowledge enabled planning that extended across generations. Writing did not merely record the past. It created the conditions for imagining the future.

The emergence of written archives marked a new cognitive landscape. Information was no longer bound by personal experience. A person could learn from records of events they never witnessed. They could inherit knowledge produced by people they would never meet. External Memory Transfer produced cumulative culture. Insights layered upon insights. Errors could be corrected. Discoveries could be preserved. The written world became an extension of the human mind, operating at a scale far beyond biology.

Yet this expansion introduced cognitive tension. Once memory moved outside the brain, individuals became reliant on external records for continuity and meaning. Written systems carried authority. They defined truth, identity, and legitimacy. A text could settle disputes or establish hierarchy. Writing created stability, but stability came with control. Those who controlled records shaped interpretation. The External Memory Transfer that freed cognition also centralized it.

Writing also reshaped internal cognition. When people relied on external records, their biological memory systems adapted. They offloaded details, focusing instead on navigation, classification, and retrieval. Memory shifted from retention to access. The mind learned to think with documents rather than through them. This adaptation strengthened abstract reasoning but weakened certain forms of internal recall. Biological memory became lighter, more flexible, and more dependent on external reference points.

In the modern world, this dependency has intensified. Digital memory systems store vast amounts of information, far exceeding the limits of the human brain. People no longer expect to remember phone numbers, addresses, directions, or even facts. They expect access. External Memory Transfer has reached a stage where memory exists primarily in devices, networks, and databases rather than in neurons. The cognitive burden has shifted almost entirely outward.

This shift creates fertile conditions for Cognitive Drift. When individuals rely heavily on external records, disruptions in those records can destabilize interpretation. Inaccurate information spreads quickly. Conflicting sources fracture shared meaning. People no longer know what to trust or how to verify. The mind becomes vulnerable to inconsistency in the external memory environment. Drift arises when the external systems that support cognition become unstable.

Understanding when memory moved outside the brain reveals a fundamental truth. Human intelligence is not defined by its biological boundaries. It is defined by the systems it creates to store, preserve, and modify knowledge. Writing marked the moment when the mind gained permanence and vulnerability in equal measure. Memory left the skull and entered the world, and in doing so, reshaped the future of cognition.

Scripts, Books, and the Longevity of Thought

When writing systems matured into scripts and books, human thought achieved a form of permanence that biology could never have provided. Early inscriptions preserved fragments of memory, but scripts transformed preservation into continuity. Books extended this continuity across centuries. This transformation represents a cognitive breakthrough known as Thought Preservation Architecture. Thought Preservation Architecture is the system through which societies stabilize ideas across time, allowing knowledge to survive far beyond the lifespan of any individual or generation.

The emergence of scripts marked the first stage of this architecture. Scripts introduced standardized forms that reduced ambiguity and increased precision. Repetition of characters created predictability. Predictability allowed thoughts to be encoded with a clarity that transcended personal voice. A scribe in Mesopotamia and a merchant in Egypt could rely on the stability of their script to communicate across distance and time. Scripts lifted thought out of its dependence on memory and placed it within a durable symbolic structure.

This precision produced a cognitive shift. Ideas could now survive intact long enough to be compared, revised, or contested. Political decrees gained authority because they remained consistent across recitations. Economic transactions became trustworthy because written records could be verified. Religious doctrines spread because their content no longer depended on the interpretation of a single storyteller. Thought Preservation Architecture created a foundation for intellectual continuity and institutional stability.

The invention of books expanded this architecture. Books consolidated multiple inscriptions into a unified cognitive container. They allowed long arguments, complex narratives, and systematic knowledge to exist as cohesive bodies of thought. A book could carry the worldview of a philosopher, the observational precision of a scientist, or the accumulated wisdom of a culture. After books appeared, no idea was limited by the capacity of a single mind or the memory of a single generation.

Books changed the nature of reading and learning. Before books, knowledge often required oral transmission, which relied on attention, presence, and memory. Books introduced private, repeatable, and self-paced engagement with ideas. A reader could return to a passage repeatedly, deepening interpretation. They could compare texts, identify contradictions, and follow reasoning across pages. Books created an environment in which thought could be examined with discipline rather than passively received. The mind became an active participant in the refinement of knowledge.

Thought Preservation Architecture grew even more powerful with the emergence of libraries. Collections of books created cognitive ecosystems. They accumulated perspectives, theories, and histories that no individual mind could contain. Libraries became external neural networks that stored collective memory. They allowed scholars to build upon centuries of accumulated insight rather than beginning from the same baseline with each generation. This cumulative acceleration reshaped the trajectory of human civilization.

However, this architecture introduced cognitive tension. The stability provided by scripts and books also created forms of rigidity. When ideas became fixed in written form, they could resist necessary adaptation. A doctrine recorded in a sacred text could constrain moral evolution. A political ideology preserved in print could limit reform. The permanence of writing strengthened societies, but it also strengthened the forces that resisted change. Thought Preservation Architecture preserved not only insight but error.

Another tension emerged from authority. Written systems conferred legitimacy. Words in a book were often treated as more accurate or trustworthy than living experience. This reverence for text made societies vulnerable to manipulation. If a written claim appeared authoritative, people accepted it even when it misrepresented reality. The architecture that preserved truth also preserved falsehood with equal durability.

In the modern era, digital systems have expanded Thought Preservation Architecture into a new domain. Knowledge is recorded instantly, replicated globally, and preserved indefinitely. Yet this amplification increases instability. When information spreads faster than verification, shared meaning fractures. Contradictory narratives coexist in the same informational space. The permanence that writing once guaranteed is now challenged by overwhelming volume and inconsistency.

These tensions create fertile ground for Cognitive Drift. Drift emerges when the external systems that preserve thought no longer align with the internal systems that interpret it. When records contradict each other, or when the pace of written information outstrips cognitive integration, the mind loses its anchoring structure. The stability that writing once provided becomes uncertain.

Understanding scripts, books, and the longevity of thought reveals how profoundly human cognition depends on external systems of preservation. Writing did not merely store knowledge. It extended the lifespan of ideas, shaped collective memory, and structured the evolution of culture. Thought Preservation Architecture is one of humanity's greatest achievements, yet it carries the seeds of its own fragility. The future of cognition will depend on whether these systems continue to stabilize thought or begin to destabilize it.

Knowledge as a Collective Inheritance

Human beings did not inherit only genes from their ancestors. They inherited ideas. They inherited methods, stories, tools, symbols, and interpretations refined over countless generations. This inheritance forms one of the most powerful forces in human evolution, a phenomenon scholars describe as Cumulative Cognitive Transmission. Cumulative Cognitive Transmission is the process through which knowledge is preserved, expanded, and transformed across time, allowing each generation to begin life with access to insights that would take individual minds centuries to discover alone.

This transmission distinguishes humans from all other species. Animals pass down instincts and sometimes simple learned behaviors, but humans inherit structured knowledge systems. These systems include techniques for agriculture, principles of navigation, moral frameworks, scientific models, and cultural narratives. Each generation receives an expanded cognitive platform on which to build. No individual begins with an empty mind. Every person enters a world already shaped by past thinkers, innovators, and storytellers.

The mechanism of Cumulative Cognitive Transmission relies on external memory systems, instructional practices, and social collaboration. Writing, ritual, apprenticeship, and storytelling all serve as transmission channels. They ensure that knowledge outlives the brain that generated it. This continuity allows societies to accumulate intellectual capital. A discovery made in one era becomes the foundation for breakthroughs in the next. A mathematical insight preserved on clay tablets contributes to celestial models developed thousands of years later. A philosophical argument recorded in a manuscript shapes ethical systems that endure to the present. Knowledge compounds, and this compounding accelerates cultural evolution.

Yet knowledge does not transfer passively. Each generation interprets and reshapes the inheritance it receives. This reinterpretation creates both growth and tension. Cumulative Cognitive Transmission preserves stability while enabling innovation. It maintains continuity while encouraging divergence. The inherited structure anchors thought, but it also invites challenge. This balance between preservation and transformation is the engine of intellectual progress.

Collective knowledge systems also distribute cognitive load. No single individual must master all the information required to maintain a complex society. Instead, knowledge becomes specialized and interconnected. Farmers inherit agricultural wisdom refined across centuries. Engineers inherit principles of mathematics and physics. Healers inherit centuries of trial, error, and empirical insight. The collective mind grows through distributed expertise. Cumulative Cognitive Transmission turns society into a cognitive organism capable of reasoning at scales that exceed the capacity of any solitary brain.

This system creates profound advantages, but it also introduces vulnerabilities. When knowledge becomes specialized, individuals must rely on experts they cannot personally verify. When knowledge becomes layered across long historical chains, errors can persist unnoticed. When knowledge becomes embedded in institutions, it can be preserved long after it ceases to be relevant. The very mechanisms that strengthen collective intelligence can weaken its adaptability.

Another tension emerges from epistemic inequality. Some individuals have greater access to knowledge than others. Some have the power to shape what is preserved and what is forgotten. Knowledge becomes a resource that can be concentrated, restricted, or weaponized. Cumulative Cognitive Transmission stabilizes societies, but it also reinforces hierarchies. The collective inheritance can unify or divide, educate or mislead, depending on how it is managed.

In the modern era, digital networks have amplified this phenomenon dramatically. Knowledge is no longer preserved through slow accumulation. It is replicated instantly and globally. The collective inheritance expands faster than individuals can comprehend. This expansion increases the potential for insight, but it also increases the potential for cognitive fragmentation. When information proliferates without structure, the shared inheritance becomes chaotic. People inherit not only wisdom but vast amounts of conflicting, unreliable, or manipulative content. The coherence of collective knowledge becomes unstable.

This instability creates fertile conditions for Cognitive Drift. Drift emerges when inherited knowledge systems lose consistency, when narratives conflict, or when individuals cannot align personal interpretation with collective information. The distributed mind that once stabilized thought becomes a source of uncertainty. The collective inheritance becomes a shifting landscape rather than a stable foundation.

Understanding knowledge as a collective inheritance reveals the deep interdependence between individual cognition and cultural evolution. Human thought advances because each generation inherits more than it could ever produce alone. Cumulative Cognitive Transmission is the scaffolding on which civilization is built. Yet this scaffolding must remain coherent and adaptable. The future of cognition will depend on how societies preserve, refine, and regulate the inheritance that has carried humanity forward for millennia.

9. Philosophy: Humanity Questions Its Own Mind

Ancient Insight Engines

Human thought crossed a transformative threshold when early philosophers began questioning the very mechanisms of the mind. Until this moment, cognition evolved primarily through biological adaptation, cultural inheritance, and technological extension. But with the appearance of reflective inquiry, humanity developed a new form of cognitive tool that did not rely on stone, symbol, or ritual. It relied on deliberate thinking about thinking. Scholars describe this breakthrough as Reflective Cognition Catalysis, the process through which philosophical reasoning accelerates the mind's ability to evaluate, refine, and reorganize its own mental structures.

The earliest philosophers operated within the myths and rituals of their cultures, yet they sensed that beneath these narratives lay deeper patterns. They asked questions that no earlier cognitive system had asked. What is truth. What is knowledge. What is the self. These questions were not abstractions. They were attempts to uncover the hidden logic guiding human perception, emotion, and understanding. By interrogating the assumptions embedded in cultural narratives, early philosophers created engines of insight that transformed cognition.

In ancient Greece, India, China, Africa, and the Middle East, thinkers independently discovered Reflective Cognition Catalysis. They shifted attention from the external world to the internal one. Observation was no longer limited to nature. It extended to thought itself. This inward turn produced conceptual categories that allowed humans to analyze reasoning, examine beliefs, and classify experience. Philosophical systems became scaffolding for new cognitive possibilities. They provided frameworks for organizing knowledge, evaluating arguments, and distinguishing illusion from reality.

These frameworks expanded the reach of human thought. In Greece, the search for first principles refined logical reasoning. In India, explorations of consciousness deepened understanding of subjective experience. In China, inquiries into harmony and order revealed the cognitive dynamics of social life. In Africa and the Middle East, traditions of ethical reflection and metaphysical interpretation shaped moral perception and narrative identity. Each tradition served as an ancient insight engine, generating models of mind that could be tested, debated, and transmitted.

Reflective Cognition Catalysis produced a form of intellectual stability that earlier societies lacked. Cultural narratives shaped meaning, but philosophical inquiry shaped interpretation. Philosophy created a meta layer above myth, allowing individuals to question the authority of inherited stories. This questioning strengthened cognition by encouraging flexibility, skepticism, and conceptual rigor. It also introduced cognitive tension. Once the mind learned it could question its own foundations, certainty became elusive. Philosophical inquiry illuminated truth but exposed ambiguity.

The creation of structured argument intensified this tension. Philosophers developed techniques such as deduction, dialectic, and systematic classification. These techniques allowed ideas to be examined with precision. They also revealed contradictions within belief systems that had seemed coherent. By exposing inconsistencies, philosophy destabilized fixed narratives and demanded new forms of coherence. Reflective Cognition Catalysis provided powerful insight, yet it also generated intellectual conflict that reshaped cultural understanding.

The influence of ancient insight engines extended far beyond their original contexts. They laid the groundwork for scientific reasoning, legal frameworks, ethical systems, and political theory. They trained the mind to separate evidence from assumption, desire from fact, and appearance from reality. Through these insights, societies developed cognitive resilience. They gained the ability to revise beliefs in response to new information. Reflective Cognition Catalysis transformed philosophy from an abstract pursuit into a tool for navigating uncertainty.

However, this mechanism carried vulnerabilities. Once societies embraced philosophical inquiry, they also embraced disagreement. Competing schools of thought emerged, each offering different interpretations of truth, selfhood, and morality. These disagreements enriched intellectual life but fractured collective meaning. The same engines that advanced cognition also created divisions. Philosophical conflict became a permanent feature of human culture.

In the modern era, this vulnerability is amplified by the sheer volume of competing ideas. Digital platforms accelerate the spread of philosophical claims without the stabilizing influence of disciplined reasoning. Reflective Cognition Catalysis becomes fragmented. People inherit partial arguments, detached insights, and conflicting worldviews. The coherence that ancient insight engines once provided becomes diluted. This dilution creates fertile ground for Cognitive Drift. Drift emerges when societies cannot maintain stable interpretive frameworks. Without shared principles of reasoning, collective understanding becomes unstable.

Understanding ancient insight engines reveals how philosophy became one of humanity's most powerful cognitive technologies. Philosophy taught the mind to reflect, question, and reorganize itself. It expanded intellectual possibility while exposing the fragility of meaning. Reflective Cognition Catalysis remains essential for navigating the complexities of modern life, yet it also reveals the challenges of maintaining coherence in an era of rapid informational transformation.

Logic, Reason, and Cognitive Discipline

Human cognition gained a new level of precision when individuals began to formalize the structure of thought. Myths organized meaning. Rituals organized memory. Philosophy organized inquiry. But logic and reason organized the mind itself. These intellectual tools created a disciplined method for evaluating ideas, distinguishing valid arguments from flawed ones, and building systems of knowledge that could withstand scrutiny. Scholars describe this transformation as Structured Rationality Formation, the process through which logic and reason shape cognition into a coherent, self-correcting system.

Structured Rationality Formation began with a recognition that thought is not naturally consistent. The mind leaps between impressions, reacts to emotion, and accepts convenient explanations without question. Early thinkers observed these tendencies and attempted to impose order upon them. They discovered that reasoning follows patterns and that these patterns can be clarified, formalized, and tested. Logic emerged as a tool for detecting contradictions. Reason emerged as a method for evaluating claims based on evidence rather than desire or tradition. Together, they became instruments for refining cognition.

The earliest logical systems provided powerful insights into the structure of thought. By examining how conclusions follow from premises, thinkers learned to differentiate valid inference from persuasive illusion. They recognized that some beliefs feel true because they align with intuition, while others feel true because they align with argument. Structured Rationality Formation created a distinction between appearance and justification. This distinction allowed the mind to correct itself. It also revealed how easily it can be misled.

As logical systems matured, they expanded beyond simple rules of inference. They introduced classification of concepts, analysis of propositions, and systematic approaches to argumentation. These tools enabled thinkers to build extensive networks of reasoning that connected diverse ideas with clarity and precision. Logic became a cognitive scaffold. It imposed discipline on thought that had previously been guided largely by narrative and emotion.

Reason amplified this discipline by introducing evaluation. Reason requires evidence, clarity, and coherence. It forces the mind to justify its claims rather than simply assert them. When combined with logic, it becomes a powerful filter that strengthens cognitive integrity. Structured Rationality Formation gave rise to rigorous debate, philosophical schools, and eventually scientific methods. It created a culture of intellectual accountability, in which ideas could be challenged, refined, or discarded based on rational criteria.

Yet this cognitive tool introduced significant tension. Logic and reason demand consistency, but lived experience is rarely consistent. Human emotions, values, and social obligations often conflict with strict rational assessment. The mind must reconcile rational discipline with the complexity of human life. This reconciliation generates cognitive strain. People may accept logical principles yet struggle to apply them when emotions dominate or social pressures intervene. Structured Rationality Formation strengthens cognition, but it also reveals the limits of rational control.

This tension deepens when rational systems become overly rigid. Logic can clarify truth, but it can also constrain imagination. Reason can correct error, but it can also suppress alternative perspectives when applied too narrowly. The demand for coherence can obscure the richness of ambiguity. Philosophical traditions that embraced strict rationality sometimes neglected the roles of emotion, intuition, and cultural context. Structured Rationality Formation provides clarity, yet clarity can become reductionism when complexity is forced into narrow frameworks.

Despite these vulnerabilities, logic and reason transformed the trajectory of human thought. They made intellectual progress cumulative and self-correcting. They laid the foundation for mathematics, science, law, and systematic ethics. They allowed societies to establish frameworks for debate and decision making that did not rely solely on authority or tradition. Structured

Rationality Formation became a cornerstone of collective intelligence.

In the modern world, this cornerstone faces new challenges. Digital environments accelerate the spread of information that bypasses rational evaluation. Emotional and symbolic content travels faster than reasoned argument. Logical coherence competes with attention-driven narratives. Structured Rationality Formation struggles to maintain influence in environments that reward speed over accuracy and emotional impact over analytical precision.

This erosion creates fertile conditions for Cognitive Drift. Drift emerges when cognitive discipline weakens and interpretive coherence dissolves. When individuals or societies no longer rely on shared rational frameworks, meaning becomes unstable. Conflicting narratives proliferate. Reason loses its authority. Without structured rationality, thought becomes vulnerable to distortion, fragmentation, and manipulation.

Understanding logic, reason, and cognitive discipline reveals their central role in shaping human intelligence. These tools refined the mind's ability to evaluate itself. They anchored knowledge in coherent principles and created standards for truth that transcend individual experience. Structured Rationality Formation remains essential for navigating complex information landscapes, yet it also requires continuous reinforcement. Without it, the stability of thought itself becomes fragile.

Building the Intellectual Architecture of Truth

Humanity's search for truth did not arise spontaneously. It emerged from the need to create reliable structures of understanding that could withstand error, bias, and the shifting pressures of culture. Individual minds are deeply influenced by emotion, memory, and social context, which makes truth precarious when left to intuition alone. To overcome these limitations, societies developed systematic methods for evaluating claims and stabilizing knowledge. Scholars describe this development as the Truth Stabilization Framework, the cognitive architecture that allows communities to construct, verify, and preserve shared models of reality.

The Truth Stabilization Framework originated from the recognition that no single perspective is sufficient for understanding the world. Early thinkers realized that observation must be tested, interpretation must be questioned, and reasoning must be evaluated. This realization created a structural shift in cognition. Truth became something that required construction rather than assumption. Philosophical inquiry, logical analysis, and emerging empirical methods formed the foundation of this architecture. They provided tools for distinguishing belief from fact and interpretation from evidence.

A key component of the Truth Stabilization Framework is methodological transparency. When thinkers describe how they reached their conclusions, others can replicate, critique, or refine their processes. This transparency transforms knowledge into a collective enterprise. It allows communities to build upon previous insights rather than relying solely on authority or tradition. The framework encourages practices such as precise definition, systematic questioning, and structured debate. These practices increase the reliability of knowledge by exposing flaws and strengthening arguments.

Another essential component is evidentiary grounding. Claims gain credibility when they are supported by observations or data that can be independently verified. This grounding protects cognition from distortion by intuition or wishful thinking. It shifts the focus from persuasion to demonstration. In early philosophical schools, this grounding appeared as careful descriptions of natural phenomena. In legal traditions, it appeared as rules of testimony. Over time, the accumulation of evidentiary methods produced increasingly rigorous standards of truth.

The Truth Stabilization Framework also requires interpretive consistency. Ideas must align with established principles unless clear justification exists for revision. This consistency prevents societies from discarding valuable knowledge due to cultural fashion or sudden emotional influence. It creates intellectual continuity. It also produces tension. When new discoveries challenge existing frameworks, societies must decide whether to preserve coherence or embrace revision. This decision shapes the trajectory of knowledge.

As the framework matured, it became institutionalized. Libraries, academies, courts, and scientific communities developed practices for storing, evaluating, and disseminating knowledge. These institutions functioned as cognitive stabilizers. They preserved reliable information while filtering out claims that lacked coherence or evidence. The Truth Stabilization Framework became embedded in social structures, influencing how societies educated their members and resolved disputes.

Yet the architecture that stabilizes truth also creates vulnerabilities. Once truth becomes institutionalized, the institutions themselves become potential sources of distortion. Power can influence which claims are preserved and which are excluded. Cultural biases can shape what counts as evidence or valid reasoning. In some periods, institutions strengthened truth by encouraging open inquiry. In others, they restricted truth by enforcing dogma. The framework that protects cognition can also constrain it.

Another tension arises from complexity. As knowledge expands, the Truth Stabilization Framework must handle increasingly sophisticated information. Interpretive systems become intricate. Standards of evidence grow more demanding. The cognitive burden of evaluating claims increases. Individuals may struggle to understand or trust the processes that produce truth. This gap between institutional reasoning and personal comprehension can weaken confidence in the framework.

In the modern era, digital information systems have amplified these tensions. Multiple, competing frameworks for truth operate simultaneously. Traditional institutions coexist with informal networks that lack rigorous standards. Information spreads faster than it can be verified. The architecture that once stabilized knowledge now competes with architectures that destabilize it. When individuals cannot discern which framework to trust, truth becomes fragmented.

This fragmentation creates fertile conditions for Cognitive Drift. Drift emerges when people can no longer rely on stable structures of interpretation. Conflicting claims, contradictory evidence, and rapidly shifting narratives overwhelm the cognitive systems that maintain coherence. The Truth Stabilization Framework weakens, and reality becomes negotiable.

Understanding the intellectual architecture of truth reveals its dual nature. It is both a triumph of human cognition and a fragile system that depends on collective discipline. The framework stabilizes thought by providing methods for evaluating and preserving knowledge, yet it remains vulnerable to distortion, complexity, and loss of trust. The future of truth will depend on whether societies can reinforce the framework in a world where information grows faster than understanding.

10. The Scientific Method: Thought Learns to Correct Itself

Observation as Revolution

Human thought entered a new epoch when people began to treat observation not as a passive experience but as a deliberate method for discovering truth. Before this shift, societies relied on mythic explanations, philosophical reasoning, and inherited authority to interpret the world. These systems provided meaning, but they did not guarantee accuracy. The cognitive leap occurred when thinkers realized that careful, systematic observation could test claims, reveal patterns, and dismantle illusions. Scholars describe this shift as Empirical Insight Activation, the process through which observation becomes a disciplined cognitive tool that reshapes understanding at its foundation.

Empirical Insight Activation began with a simple realization. The world offers signals that can be examined directly. These signals are not dependent on belief, cultural narrative, or personal preference. They exist whether or not they fit human expectations. Early natural philosophers discovered that when they paid close attention to these signals, they often contradicted accepted explanations. The movement of planets differed from mythic cosmology. The behavior of light contradicted intuitive models. Biological processes resisted explanation by symbolic analogy. Observation revealed that reality does not bend to narrative.

This realization introduced a dramatic cognitive tension. If observation could contradict tradition, then tradition could no longer serve as the final arbiter of truth. Empirical Insight Activation challenged the interpretive authority of myth, ritual, and intuition. It placed evidence above belief. This priority created a method for correcting error. Thought could revise itself based on new information rather than preserving outdated narratives. Observation became a force that disciplined cognition.

To harness this power, thinkers developed systematic methods for observing natural phenomena. They recorded details, measured quantities, and repeated investigations to confirm results. Observation became structured rather than spontaneous. It demanded patience, clarity, and precision. Through this discipline, Empirical Insight Activation produced insights that no earlier cognitive system could generate. People began to understand motion, energy, anatomy, and chemical transformation. They discovered patterns that underlay the apparent chaos of nature. The mind learned that the world is intelligible if approached with the right tools.

This transformation reshaped the relationship between perception and understanding. In earlier cognitive systems, perception was filtered through mythic or symbolic frameworks. Observation served interpretation. In the scientific method, interpretation served observation. This reversal elevated perception into a cognitive engine rather than a passive receiver. Empirical Insight Activation allowed perception to challenge expectation. It empowered the mind to interrogate itself by comparing belief against evidence.

Yet this empowerment introduced new vulnerability. Observation is not purely objective. It is shaped by instruments, techniques, and cognitive biases. Early scientists quickly learned that the mind often sees what it expects to see. Instruments could distort results. Conditions could mislead. Empirical Insight Activation required not only careful observation but careful skepticism. The mind needed methods to prevent itself from being deceived by its own interpretations. This requirement led to replication, peer evaluation, and methodological transparency. The scientific method emerged as a disciplined extension of the mind's natural curiosity.

As observation became central to knowledge, societies developed new forms of authority. Expertise shifted from storytellers and philosophers to investigators who could demonstrate claims through evidence. This shift created a new hierarchy of trust. It also produced tension. When evidence contradicted deeply held beliefs, individuals struggled to integrate the new information. The scientific method provided clarity, but clarity often clashed with cultural or emotional commitments. Empirical Insight Activation strengthened cognition while provoking conflict between evolving truth and inherited meaning.

In the modern world, observation remains revolutionary. Technologies such as telescopes, microscopes, sensors, and simulations extend the reach of perception far beyond biological limits. These extensions allow humanity to observe galaxies, genomes, and quantum systems. Yet each extension also increases complexity. Observers must interpret data filtered through layers of instruments and algorithms. Empirical Insight Activation becomes vulnerable to misinterpretation, overconfidence in models, or manipulation of evidence. The revolutionary clarity of observation coexists with the risk of sophisticated error.

This risk creates fertile conditions for Cognitive Drift. Drift emerges when individuals or societies rely on evidence they cannot evaluate, or when conflicting observations undermine coherent interpretation. Digital environments amplify this problem. Observational claims circulate rapidly, without context or verification. The clarity once provided by empirical investigation becomes clouded by informational noise. When observation loses its stabilizing function, the cognitive system that depends on it becomes fragile.

Understanding observation as revolution reveals the profound transformation that occurs when thought learns to correct itself. Empirical Insight Activation freed cognition from inherited constraints and allowed truth to emerge through disciplined attention to the world. It remains one of humanity's most powerful cognitive tools, but it also requires vigilance. Observation can illuminate reality, but only when guided by structures that prevent distortion. The future of knowledge will depend on whether societies can preserve the integrity of this revolutionary method.

Models, Evidence, and Cognitive Precision

The scientific method did more than elevate observation. It introduced a new structure for organizing thought itself. Humans have always formed mental models, but before science, these models were rarely tested with disciplined rigor. They emerged from intuition, tradition, or narrative convenience. The scientific revolution transformed this landscape by creating systematic methods for building, evaluating, and refining models of reality. Scholars refer to this transformation as the Precision Modeling Framework, the cognitive system through which hypotheses are constructed, tested against evidence, and repeatedly corrected to increase accuracy.

The Precision Modeling Framework rests on a simple insight. The mind does not perceive the world directly. It constructs internal representations that approximate external reality. These representations guide action, but they can be mistaken. To correct them, thinkers must compare their models with observable data and measure the gap between prediction and outcome. This comparison transforms cognition into a dynamic system that improves through iteration. The mind becomes an instrument of self-correction.

Early scientists recognized that models require clarity. They must be explicit enough to be tested and structured enough to be falsified. This requirement introduced a discipline that earlier forms of reasoning lacked. Models could no longer rely on vague generalities or poetic metaphors. They needed precision in definition and logical structure. The Precision Modeling Framework demanded that claims be framed in ways that allowed others to examine and challenge them. This demand deepened cognitive accountability.

Evidence became the central arbiter in this framework. It functioned as an external checkpoint that grounded models in the real world. Thinkers measured, recorded, and compared data, seeking patterns that confirmed or contradicted their hypotheses.

The role of evidence extended beyond confirmation. It taught thinkers to accept uncertainty and revise assumptions. The framework encouraged humility. No model, however elegant, could stand without empirical support. The authority of thought shifted from intuition to verification.

This shift created cognitive resilience. Models that survived repeated testing gained reliability. Those that failed were refined or discarded. Over time, the Precision Modeling Framework generated structures of knowledge with unprecedented accuracy. Newton's laws, anatomical studies, chemical classifications, and biological taxonomies emerged from continuous cycles of modeling and correction. Each cycle expanded humanity's ability to predict and manipulate natural phenomena. Precision became power.

Yet precision introduced tension. The more detailed a model became, the more complex the evidence required to support it. Instruments grew sophisticated. Data sets expanded. Interpretation became increasingly dependent on specialized knowledge. This specialization created distance between experts and the public. Many individuals relied on models they could not personally evaluate. The Precision Modeling Framework strengthened knowledge while weakening direct comprehension.

The tension deepened as models became abstract. Theories in physics, biology, and economics often described phenomena far removed from everyday experience. Their validity rested on mathematical coherence and experimental confirmation that few could replicate. This abstraction increased the cognitive gap between model makers and model users. Societies gained advanced predictive tools but lost intuitive grounding in the reasoning behind them. Trust became essential, yet trust remained vulnerable.

Modern technology intensified these challenges. Digital systems generate enormous volumes of data and produce models that operate at scales beyond human attention. Algorithms detect patterns invisible to unaided analysis. Simulation replaces direct observation. The Precision Modeling Framework now depends on tools that mediate perception so heavily that even experts may struggle to understand how conclusions are produced. Precision grows, but transparency diminishes.

This diminishing transparency creates fertile conditions for Cognitive Drift. Drift emerges when individuals or societies rely on models without understanding their assumptions or limitations. Conflicting models proliferate. Evidence becomes contested. Data can be manipulated or selectively interpreted. As coherence weakens, people gravitate toward explanations that match emotional or cultural expectations rather than empirical grounding. The framework that once stabilized knowledge becomes a site of fragmentation.

Understanding models, evidence, and cognitive precision reveals both the genius and fragility of scientific reasoning. The Precision Modeling Framework allows thought to correct itself systematically. It transforms the mind from a narrative-driven interpreter into an evidence-driven analyst. It creates predictive accuracy and intellectual resilience. Yet this framework depends on trust, transparency, and shared standards of evaluation. Without these stabilizers, even the most precise models lose their cognitive anchoring.

The future of scientific thought will depend on reinforcing the structures that maintain precision. The mind must continue refining its models while ensuring that the systems supporting those models remain coherent, trustworthy, and comprehensible. Only then can the Precision Modeling Framework fulfill its promise as humanity's most powerful method for understanding reality.

The Age of Organized Curiosity

Human curiosity existed long before science, but it was scattered and inconsistent. It flickered through moments of wonder without a stable structure to guide it. The scientific revolution transformed curiosity from a personal impulse into a collective engine of discovery. This transformation created a new cognitive system that scholars describe as Structured Curiosity Networks. Structured Curiosity Networks refer to the institutions, practices, and collaborative methods that channel individual curiosity into organized, cumulative inquiry. Through this system, humanity gained the ability to pursue questions with precision, coordination, and persistence across generations.

Before these networks emerged, curiosity was fragile. It could be suppressed by cultural norms, overwhelmed by daily survival, or distorted by mythic interpretation. Individuals investigated only what they found personally intriguing. Their discoveries rarely connected with the insights of others. Knowledge grew sporadically, advancing in isolated bursts that often vanished when the discoverer died. Structured Curiosity Networks changed this dynamic by externalizing curiosity into shared practices and institutional frameworks. Inquiry became a social project rather than an individual pursuit.

The first step in this transformation was the development of collaborative investigation. Scientists began to share observations, criticize each other's interpretations, and refine theories through debate. Curiosity became communicative. A question posed by one thinker could be pursued by another. A failure encountered in one laboratory could become a catalyst for success in another. Structured Curiosity Networks created feedback loops that amplified investigation. Each discovery generated new questions, and each question attracted new minds.

The second step was the establishment of formal institutions dedicated to inquiry. Academies, universities, observatories, and scientific societies emerged as stable environments for investigation. These institutions stored knowledge, standardized methods, and cultivated generations of thinkers. They protected curiosity from political instability and economic pressure by granting it legitimacy and resources. The stability of these structures allowed curiosity to expand its scope, tackling problems that required decades of sustained attention.

The third step involved the creation of methodological norms. Peer review, replication, controlled experimentation, and standardized measurement transformed curiosity into a disciplined activity. These norms ensured that inquiry produced reliable knowledge rather than persuasive speculation. Structured Curiosity Networks imposed a cognitive discipline that strengthened scientific integrity. Curiosity no longer wandered aimlessly. It followed trails of evidence, corrected itself when necessary, and integrated diverse insights into coherent models.

This transformation had profound cognitive consequences. It allowed humanity to uncover phenomena that intuition could never reveal. Evolution, electromagnetism, germ theory, and quantum mechanics emerged not from individual imagination but from collective inquiry. Structured Curiosity Networks expanded the cognitive horizon of the species. They made it possible to explore scales of reality far beyond human perception, from subatomic particles to distant galaxies.

Yet this system introduced tension. Curiosity thrives on openness, but structured inquiry requires rules. The balance between freedom and discipline is delicate. Too much structure can restrict exploration, discouraging unconventional ideas. Too little structure can produce confusion, error, and fragmentation. Structured Curiosity Networks must constantly adjust to maintain this balance. When they succeed, knowledge accelerates. When they fail, inquiry stagnates.

Another tension emerges from specialization. As scientific knowledge expands, researchers focus on increasingly narrow domains. This specialization increases precision but reduces shared understanding. Scientists in different fields may struggle to communicate, even when their discoveries are interconnected. The network grows powerful but fragmented. Curiosity becomes organized, yet compartmentalized. The collective mind gains depth at the cost of unity.

Modern technology intensifies these tensions. Digital tools accelerate data collection and analysis, but they also flood researchers with information. Algorithms generate hypotheses faster than humans can interpret them. Global networks enable collaboration but also create competition for attention. Structured Curiosity Networks expand rapidly, yet the coherence of inquiry becomes more difficult to maintain. Unverified claims circulate alongside verified ones. Communication becomes instantaneous, but verification becomes strained.

These conditions create fertile ground for Cognitive Drift. Drift emerges when the systems that stabilize inquiry weaken. When evidence is overwhelmed by volume, when institutions lose trust, or when competing interpretations proliferate without resolution, the cognitive framework that once guided curiosity begins to falter. Organized curiosity collapses into scattered investigations that lack shared standards. The collective pursuit of truth dissolves into informational noise.

Understanding the Age of Organized Curiosity reveals how humanity transformed a primal impulse into a disciplined engine of discovery. Curiosity became cumulative, collaborative, and self-correcting. Structured Curiosity Networks allowed the species to expand its cognitive reach beyond imagination. Yet these networks depend on coherence, trust, and methodological rigor. If these foundations weaken, the system that once anchored thought may become a source of instability.

The future of knowledge will depend on whether societies can preserve the structures that keep curiosity organized, disciplined, and aligned with truth.

PART - IV
THE MODERN COGNITIVE LANDSCAPE
THE HUMAN MIND UNDER NEW PRESSURES

11. The Age of Psychology: Mapping the Invisible

Behavior, Mind, and Mechanisms

The scientific study of psychology began with a radical recognition. Human behavior could be examined with the same rigor used to study physical systems. Thought, emotion, perception, and action were not mysteries reserved for philosophy. They were measurable patterns produced by mechanisms inside the brain and expressed through observable behavior. This recognition created a turning point in our understanding of the mind, a development known as Mechanistic Psychological Mapping. Mechanistic Psychological Mapping is the systematic attempt to infer internal cognitive processes from patterns of behavior and physiological signatures.

Before this shift, explanations of behavior relied largely on introspection, mythic reasoning, or moral interpretation. People acted because of will, spirit, temperament, or divine influence. These explanations were powerful but imprecise. Psychology emerged when thinkers proposed that behavior follows rules that can be discovered, tested, and revised. They observed that humans respond to incentives, form habits, learn from consequences, and exhibit patterns that repeat across environments. These patterns revealed that the mind is not a vague container of experiences. It is a structured system that transforms stimuli into action.

Mechanistic Psychological Mapping began with the study of observable behavior. Early researchers identified consistent relationships between stimuli and responses. Habits formed through repetition. Learning occurred through reinforcement.

Attention shifted according to salience. These discoveries grounded psychological science in empirical observation. They demonstrated that behavior is shaped by underlying processes, even when those processes remain hidden from awareness. The mind became a system that could be studied indirectly through its outputs.

Yet behavior alone could not explain the full richness of human cognition. Researchers soon recognized that internal mechanisms mediate experience in ways that cannot be reduced to external actions. Memory is selective. Perception is filtered. Emotion modifies interpretation. Thought does not merely react to the world. It constructs meaning. Mechanistic Psychological Mapping expanded to include internal cognition, creating a layered model in which behavior, neural activity, and subjective experience form interconnected systems.

This expansion introduced a powerful form of cognitive analysis. By comparing behavior across contexts, psychologists could infer the mechanisms that generate it. By observing reaction times, they could estimate cognitive load. By tracking eye movements, they could map attention. By measuring physiological responses, they could detect emotion beneath language. These tools revealed that much of human thinking occurs outside conscious awareness. Consciousness narrates experience, but mechanisms drive it.

This discovery created profound cognitive tension. People believe they act from intention, autonomy, and rational decision making. Mechanistic Psychological Mapping showed that behavior is influenced by biases, heuristics, and subconscious processes. These findings challenged traditional assumptions about free will and personal identity. They revealed that the mind is less transparent to itself than most people imagine. Psychological science demonstrated that individuals often misunderstand the causes of their own actions.

The rise of cognitive psychology deepened this tension. Researchers uncovered systematic errors in reasoning, distortions in memory, and predictable patterns of misjudgment. The mind uses shortcuts to navigate complexity, but these shortcuts can mislead. Mechanistic Psychological Mapping exposed the fragility of intuition and the unreliability of introspection. These insights forced a re-evaluation of how people understand themselves and others. The mind became both remarkable in its adaptability and vulnerable in its inconsistency.

Modern psychology integrates behavior, cognition, and neural mechanisms. It examines how networks of neurons produce attention, how memory encoding shapes identity, and how emotional circuits guide decision making. Mechanistic Psychological Mapping now extends into fields such as behavioral economics, computational psychiatry, and cognitive neuroscience. The mind is no longer a single entity. It is a layered system of interacting mechanisms that operate at different speeds and levels of awareness.

These advances reveal new vulnerabilities. As psychological knowledge expands, so does the ability to manipulate behavior. Advertising exploits attention. Social media platforms amplify reward loops. Political messaging targets cognitive biases. Mechanistic Psychological Mapping gives societies insight into the mind, but it also gives institutions the power to shape it. This power creates fertile conditions for Cognitive Drift. Drift emerges when external pressures distort internal mechanisms, causing shifts in identity, perception, or belief that the individual cannot trace or explain.

Understanding behavior, mind, and mechanisms clarifies the stakes of modern cognition. Psychology made the invisible visible. It mapped the processes that shape thought and revealed the forces that undermine it. Mechanistic Psychological Mapping remains one of humanity's most powerful tools for understanding itself, yet it also exposes the fragility of the cognitive systems we depend on.

Bias, Heuristics, and Cognitive Limits

The human brain did not evolve for perfect reasoning. It evolved for rapid decision making in environments filled with uncertainty, danger, and incomplete information. To navigate such environments, the mind developed shortcuts that simplified complex problems into manageable choices. These shortcuts, known as heuristics, provide speed and efficiency at the cost of accuracy. They are products of a cognitive principle termed Adaptive Efficiency Biasing. Adaptive Efficiency Biasing refers to the tendency of the mind to favor rapid, energy saving judgments that prioritize survival over precision.

Heuristics were essential for early humans. A quick inference about a rustling sound could mean the difference between life and death. A fast assumption about trustworthiness could determine the success of cooperation. In such contexts, a rapid but imperfect judgment offered greater survival value than slow, analytical reasoning. Adaptive Efficiency Biasing shaped neural architecture, producing cognitive systems that favor patterns, salience, and emotional cues. These systems remain active today, influencing decisions even in environments where the risks no longer resemble ancestral threats.

However, heuristics come with cognitive costs. They create predictable patterns of error known as biases. Biases are not flaws in the moral sense. They are by-products of the brain's attempts to conserve energy, simplify information, and maintain coherence under pressure. Confirmation bias, for example, reinforces existing beliefs because searching for contradictory evidence requires more cognitive effort. Availability bias makes rare events seem common when they are vivid or emotionally charged. Anchoring bias shapes judgment by clinging to the first piece of information encountered. Each bias reveals how the mind defaults to efficiency rather than accuracy.

Adaptive Efficiency Biasing introduces a sharp tension between perception and reality. The cognitive shortcuts that once protected early humans now distort decision making in modern environments. The world contains far more information than the brain can process. Media intensifies emotional stimuli.

Technology accelerates the pace of interaction. Institutions frame choices in ways that exploit cognitive shortcuts. The mind struggles to maintain coherence. It relies on heuristics even when precision is required, producing errors that feel intuitive but are fundamentally flawed.

These biases affect not only decisions but identity. People build self-narratives that rely on selective attention, motivated reasoning, and memory distortions. The mind protects the self by shaping interpretation in ways that maintain emotional stability. This protection can become a barrier to growth. Adaptive Efficiency Biasing prevents individuals from recognizing inconsistencies, adjusting beliefs, or confronting uncomfortable truths. It stabilizes the self at the cost of accuracy.

The study of cognitive limits revealed another dimension. Human working memory is narrow. Attention is easily disrupted. Complex reasoning is fragile. People overestimate their capacity to process information, especially in environments flooded with stimuli. These limits interact with heuristics in ways that amplify error. When cognitive load increases, people rely more heavily on shortcuts. When attention fragments, bias strengthens. Adaptive Efficiency Biasing becomes more dominant under stress, fatigue, or ambiguity.

Modern life magnifies these weaknesses. Digital platforms exploit biases by presenting emotionally charged information that captures attention. Algorithms amplify content that triggers heuristic responses. Social groups reinforce shared narratives, strengthening confirmation bias. Political messaging frames issues in ways that exploit cognitive shortcuts. Adaptive Efficiency Biasing becomes a point of vulnerability rather than a survival tool.

These vulnerabilities create fertile ground for Cognitive Drift. Drift emerges when biases become strong enough to alter a person's interpretive framework. Small distortions accumulate. Selective attention redirects perception. Memory becomes reconstructed according to emotional needs. Over time, individuals may enter cognitive environments that diverge from reality without realizing the process. Adaptive Efficiency Biasing becomes unstable under informational overload, allowing Drift to reshape identity and belief.

Understanding bias, heuristics, and cognitive limits reveals the fundamental nature of human thought. The mind is not an objective machine. It is an adaptive engine shaped by evolutionary pressures that prioritized survival over truth. These adaptations remain powerful, but they operate in environments for which they were never designed. The challenge of modern cognition is to recognize these limits and develop structures that compensate for them. Only by understanding Adaptive Efficiency Biasing can societies build safeguards that protect individuals from systematic distortion, fragmentation, and Drift.

The Hidden Architecture of Emotion

Emotion is often treated as the opposite of reason, a force that disrupts clarity and distorts judgment. Yet emotion is not a deviation from cognition. It is a core component of the mental system, an ancient architecture that guides perception, prioritizes information, shapes memory, and directs behavior. The evolutionary purpose of this architecture is described by scholars as Affective Guidance Infrastructure. Affective Guidance Infrastructure refers to the network of neural and physiological mechanisms that assign value to experience and orchestrate the body's response long before conscious thought intervenes.

In early evolutionary environments, survival depended on rapid interpretation of threats and opportunities. Emotion provided this interpretive speed. Fear accelerated attention and mobilized energy. Curiosity encouraged exploration. Affection strengthened bonds essential for group living. These emotional states were not irrational impulses. They were adaptive regulators of behavior. Affective Guidance Infrastructure translated raw sensory input into motivational signals that increased the likelihood of survival. Without emotion, early organisms could not have navigated a world filled with uncertainty.

The architecture of emotion continued to evolve as brain complexity increased. It integrated with memory, allowing organisms to assign significance to past events. Emotion highlighted which experiences mattered and which could be forgotten. A negative event became a warning. A pleasurable one became a guide. Emotional salience shaped the contours of memory, determining how experience influenced future behavior. This integration created a powerful feedback loop. Emotion shaped memory, and memory reinforced emotion.

Affective Guidance Infrastructure also shaped perception. People rarely see the world neutrally. They see it through emotional filters. Threat increases the detection of danger. Joy broadens attention. Sadness narrows focus. Emotion influences which stimuli rise to awareness and how they are interpreted. This influence can produce clarity when conditions align with evolutionary expectations. It can also produce distortion when modern environments evoke emotions disproportionate to actual risk.

The architecture extends into social cognition. Humans rely on emotional cues to understand the intentions of others. Facial expressions, tone of voice, and posture communicate information about cooperation, conflict, trust, or threat. Emotional resonance strengthens group cohesion and supports collective action. Without these mechanisms, societies could not maintain stability. Affective Guidance Infrastructure creates synchrony between individuals, allowing shared experiences to shape group identity and cultural norms.

Yet this architecture introduces deep cognitive tension. Emotion is essential for adaptive behavior, but it can override rational evaluation. In ancestral environments, emotional responses aligned closely with physical realities. In the modern world, abstract information, mediated communication, and symbolic threats activate emotions that evolved for direct, immediate danger. A fearful headline can trigger the same physiological response as a predator. Social rejection can feel like physical harm. This mismatch between ancient emotional mechanisms and modern stimuli creates distortions in judgment.

Emotion also competes with cognitive control. The prefrontal regions responsible for reasoning often struggle to regulate emotional responses when intensity is high. Under stress, people revert to instinctive reactions. Affective Guidance Infrastructure overwhelms analytical thinking. This competition shapes moral decisions, risk assessment, and interpersonal conflict. Cognitive control is fragile compared to the speed and force of emotional activation.

Another tension arises from emotional contagion. In groups, emotions spread rapidly. A single act of fear can cascade into collective panic. A message filled with anger can reshape an entire community's perception of reality. Affective Guidance Infrastructure evolved to synchronize group behavior during crises, but in modern societies, it creates vulnerability to manipulation. Media, political rhetoric, and digital networks can trigger large scale emotional shifts detached from actual conditions. The architecture that once stabilized group life can destabilize it.

These dynamics create fertile ground for Cognitive Drift. Drift emerges when emotional bias reshapes memory, perception, or belief in ways that break coherence. A person may reinterpret past events according to present emotion, altering their identity narrative. Groups may adopt shared emotional interpretations that diverge from evidence. Affective Guidance Infrastructure becomes a driver of Drift when emotional intensity overwhelms cognitive integration.

Understanding the hidden architecture of emotion reveals that feelings are not obstacles to thought. They are engines of meaning, motivation, and connection. They guide attention, shape memory, and regulate social life. Yet this architecture requires balance. Emotion must inform cognition without dominating it. Affective Guidance Infrastructure is powerful enough to anchor or destabilize the mind. The future of coherent thought depends on how well individuals and societies manage the emotional systems that have shaped humanity from its earliest beginnings.

12. The Digital Mind: How Technology Rewired Us

Attention Under Siege

Human attention is one of the most valuable cognitive resources ever produced by evolution. It determines what enters consciousness, what becomes memory, and what guides behavior. For millions of years, attention was shaped by natural pressures. It focused on danger, opportunity, social signals, and meaningful patterns in the environment. But in the digital era, this ancient system encounters a radically different landscape. The environment no longer competes for attention organically. It is engineered to capture it. This transformation is known as Attention Capture Hyper competition, the accelerated contest in which digital systems, platforms, and algorithms fight for control of human focus.

Attention Capture Hyper competition began with simple interfaces that offered communication and information. But as digital systems evolved, attention became the currency of technological success. Platforms optimized for engagement began using design strategies meant to hijack the architecture of attention. Notifications triggered urgency. Infinite scrolling eliminated natural stopping points. Algorithms predicted which stimuli would produce the strongest reaction. Each feature was designed to maximize the amount of time a user remained engaged. The goal was not to inform or enrich. It was to extract attention.

The cognitive system that evolved to track predators and interpret social cues was unprepared for this environment. A human can resist a single distraction. They cannot resist a system engineered to exploit every bias, emotion, and vulnerability. Digital platforms combine visual salience, novelty, unpredictability, and social validation in ways that trigger ancient attentional reflexes. Attention Capture Hyper competition overwhelms the mind because it taps into mechanisms designed for survival, not digital stimulation.

This shift produces significant cognitive tension. Attention is finite, but digital environments demand infinite engagement. The mind attempts to adapt by fragmenting. It divides focus across multiple streams of information, switching rapidly between tasks. This phenomenon, known as attentional fragmentation, weakens cognitive endurance, reduces working memory capacity, and disrupts deep thinking. The mind becomes oriented toward immediacy rather than sustained analysis. Thought loses continuity.

Neuroscientific research shows that fragmented attention alters the brain's functional networks. Pathways associated with reward become more active, while pathways associated with long term concentration weaken. The individual experiences a constant pull toward stimulation. The ability to maintain focus without external reinforcement declines. The digital mind becomes reactive rather than reflective. Attention Capture Hyper competition reshapes cognition at both structural and phenomenological levels.

This transformation affects emotional life as well. Digital environments amplify stimuli that evoke strong affective responses because such stimuli increase engagement. Content that provokes outrage, fear, admiration, or tribal belonging spreads more rapidly. These emotional triggers further hijack attention. People become caught in loops of compulsive checking, seeking emotional cues that reinforce identity or provide momentary validation. The mind shifts from autonomous focus toward stimulus driven reactivity.

The social consequences are profound. Groups exposed to different information streams develop divergent attentional worlds. They attend to different threats, narratives, and values. Without shared attention, shared reality weakens. Collective understanding becomes fractured. Communities lose the common cognitive ground necessary for cooperation. Attention Capture Hyper competition becomes a driver of social fragmentation.

These dynamics create fertile conditions for Cognitive Drift. Drift emerges when attentional instability disrupts memory consolidation, distorts interpretation, or fractures continuity of thought. When attention cannot settle long enough to integrate experience, the mind loses coherence. Thoughts appear disconnected. Identity becomes fluid. The narrative self destabilizes. Drift is not simply a failure of memory. It is a failure of attention, the gatekeeper of cognition.

Understanding attention under siege reveals the fragility of human focus in an engineered environment. The digital world does not merely distract. It restructures the mechanisms that govern awareness. Attention Capture Hyper competition transforms the mind into an arena where external systems compete for cognitive control. This competition threatens sustained intelligence, emotional regulation, and collective stability.

The challenge of the future is not only to manage information but to reclaim attention as a cognitive resource. Human survival once depended on the ability to detect the faintest rustle in the dark. Today it depends on the ability to resist engineered stimuli that overwhelm the ancient architecture of focus. The next evolution of thought will require systems that protect attention rather than exploit it.

Memory in the Cloud

Human memory evolved to store fragments of experience, patterns of survival, and socially transmitted knowledge. For most of human history, memory lived inside the brain, supported by oral tradition, written records, and collective practices. But digital technology has relocated memory once again, this time into vast external infrastructures. This transition represents a new cognitive phenomenon known as Cloud Memory Offloading. Cloud Memory Offloading refers to the process by which individuals outsource storage, retrieval, and organization of information to digital systems that operate at scales beyond biological capability.

The shift began when humans started storing photographs, messages, schedules, and documents in digital devices. Over time, these systems grew more integrated, more searchable, and more omnipresent. Memory no longer depended on personal effort. With a single query, a person could retrieve facts, images, or conversations from years earlier. The brain adapted by offloading details that once required attention and rehearsal. Cognitive energy redirected from retention to navigation. People learned how to access information rather than store it.

Cloud Memory Offloading expanded further with the rise of predictive systems. Devices began anticipating the needs of their users. Calendars reminded individuals of tasks. Applications resurfaced forgotten memories. Algorithms suggested information based on patterns of behavior. Memory became not only externalized but automated. The digital environment began shaping which memories resurfaced and when. The individual no longer controlled retrieval. It became a shared activity between mind and machine.

This transformation created new cognitive strengths. External memory systems offered precision, permanence, and efficiency. They preserved enormous quantities of information without decay. They allowed individuals to operate in complex environments without overwhelming working memory. Cloud Memory Offloading made it possible to manage schedules, relationships, and knowledge at a scale impossible for biological memory alone.

Yet this shift introduced profound tension. Biological memory is selective for a reason. It compresses experience into meaningful patterns. It strengthens the self by reinforcing continuity. Digital memory does not discriminate. It preserves trivial details alongside significant ones. It surfaces memories without considering emotional readiness or relevance. This mismatch destabilizes the natural rhythm of remembering and forgetting. The mind becomes dependent on systems that do not share its priorities.

Another tension arises from vulnerability. Biological memory is internal and secure. Cloud memory is external and fragile. Data can be lost, altered, or accessed without permission. The individual develops reliance on infrastructures they cannot control. This reliance diminishes cognitive independence. When external memory fails, people experience disorientation, anxiety, or functional collapse. Cloud Memory Offloading strengthens cognition while weakening autonomy.

A deeper consequence involves identity. Personal identity is shaped through memory integration, the process by which the mind organizes experiences into a coherent narrative. When memories are stored externally, this integration becomes fragmented. A person may revisit digital archives that contradict their internal narrative, destabilizing the story they have constructed about themselves. The continuity of identity becomes dependent on platforms that curate which memories reappear.

These vulnerabilities create fertile conditions for Cognitive Drift. Drift emerges when external memory systems overwrite or disrupt the internal narrative. Automated reminders may evoke emotions detached from present context. Algorithmic resurfacing of old images may distort interpretation of past relationships. Cloud Memory Offloading becomes a source of instability when it shapes remembering in ways that bypass conscious control. The mind loses coherence as external systems influence which memories feel vivid or significant.

Understanding memory in the cloud reveals the dual nature of digital memory systems. They extend cognitive capacity while altering the structure of experience. They preserve detail while destabilizing meaning. Cloud Memory Offloading represents a new phase in the evolution of thought, one in which the boundary between internal and external memory blurs. The challenge is not to reject these systems but to develop practices that maintain autonomy, coherence, and emotional stability in an environment where machines increasingly decide what we remember.

The Algorithmic Self

Human identity once formed through introspection, social feedback, and cultural narratives. Individuals constructed a sense of self by interpreting their experiences, relationships, and aspirations. But in the digital era, a new force participates in this construction. Algorithms observe behavior, predict preferences, and shape digital environments that reflect those predictions. This interaction creates a cognitive phenomenon called Synthetic Identity Formation. Synthetic Identity Formation refers to the process by which algorithms influence, reinforce, or reshape an individual's sense of self through personalized digital feedback loops.

The emergence of this process began with simple recommendation systems. Platforms tracked clicks, searches, and preferences to suggest content. These suggestions altered behavior, which generated new data that strengthened the algorithm's model of the user. Over time, these systems grew more sophisticated. They learned to anticipate desires, guide attention, and personalize digital environments. The individual encountered versions of reality tailored to their inferred identity.

This personalization is not passive. It is participatory. The behavior of the user shapes the model, and the model shapes the behavior of the user. This reciprocal dynamic turns identity into a computational feedback loop. Synthetic Identity Formation emerges from the interplay between human intention and algorithmic prediction. The self becomes a hybrid construction, part biological, part digital, continuously recalibrated by systems that measure and respond to behavior with extraordinary speed.

This transformation introduces unprecedented psychological tension. Identity becomes less a product of deliberate reflection and more a product of algorithmic inference. People begin to internalize the preferences that algorithms amplify. They adopt behaviors that align with predicted patterns. They see information that confirms their modeled identity rather than information that challenges it. Over time, the algorithmically curated environment narrows the self into a stable pattern optimized for engagement rather than growth.

This narrowing affects autonomy. Traditional identity development involves exploration, contradiction, and uncertainty. Synthetic Identity Formation reduces exposure to unfamiliar perspectives. It rewards predictability. The individual becomes a set of behavioral probabilities rather than a dynamic psychological process. Authentic decision making becomes entangled with algorithmic nudging. Freedom of thought weakens as the digital environment constrains the range of possible selves.

The social consequences deepen this tension. Individuals shaped by different algorithmic ecosystems develop divergent worldviews, values, and emotional responses. Their online environments reinforce distinct identities that rarely intersect. Synthetic Identity Formation becomes a driver of polarization. People inhabit separate cognitive realities even while sharing physical space. The collective mind fractures into algorithmically sorted clusters.

Another vulnerability emerges from invisibility. Algorithms operate behind interfaces that conceal their influence. People rarely understand how their preferences are inferred or how their identities are shaped. This opacity makes Synthetic Identity Formation difficult to resist. The self-drifts gradually, shaped by patterns that feel natural but originate from external systems. Identity becomes porous to forces that do not prioritize psychological well-being or coherence.

These conditions create fertile ground for Cognitive Drift. Drift arises when the algorithmic environment reshapes interpretation, emotion, or memory in ways the individual cannot trace. A person may feel beliefs strengthening without understanding why. They may adopt values that align with algorithmic predictions rather than personal conviction. They may experience shifts in identity triggered by patterns of interaction rather than self-reflection. Synthetic Identity Formation becomes a mechanism of Drift when the feedback loop diverges from the individual's authentic cognitive trajectory.

Understanding the algorithmic self-reveals a profound transformation in human identity. The self is no longer defined solely by introspection, culture, or social interaction. It is shaped by invisible systems that predict and influence behavior. Synthetic Identity Formation marks a new era in cognition, one in which autonomy depends on recognizing and managing the algorithmic forces that shape experience. The future of identity will depend on whether individuals can reclaim agency within digital environments that constantly reconstruct who they are.

13. The Global Brain

Networks as Cognitive Ecosystems

Humanity has always relied on connection. Families formed tribes. Tribes formed communities. Communities formed nations. But the digital era introduced a new scale of connection, one that links billions of minds into a single, ever-shifting web of communication. This web does more than transmit information. It shapes thought itself. Scholars refer to this transformation as Networked Cognition Ecology. Networked Cognition Ecology describes the way digital networks function as living cognitive ecosystems that process information, regulate attention, and influence belief across vast populations.

The foundation of this ecology lies in connectivity. When individuals interact through digital platforms, their thoughts, emotions, and interpretations circulate through shared channels. Each message, image, and idea becomes a cognitive signal that influences the perceptions of others. These signals propagate through the network, creating patterns that resemble neural activity. Clusters of people form cognitive nodes. Their interactions form pathways. Trends and conversations act as pulses of activation that ripple across regions of the digital world. The network behaves like an external mind with its own dynamics, feedback loops, and emergent properties.

This transformation alters the scale of human cognition. An individual no longer thinks in isolation. Their thoughts are influenced by global flows of information that travel at speeds the biological brain cannot match. Networked Cognition Ecology amplifies the reach of every idea, regardless of its accuracy. A rumor can circle the planet within minutes. A scientific insight can inspire global collaboration. A single emotional expression can trigger collective outrage or collective hope. The network accelerates cognition, but it does not discriminate between clarity and distortion.

The architecture of digital networks introduces sharp cognitive tension. Biological cognition evolved in small groups where information could be evaluated through personal experience, shared norms, and stable relationships. In contrast, digital ecosystems present information with little context, little continuity, and little accountability. The mind is not adapted to receive a constant stream of claims from unknown sources. Networked Cognition Ecology overwhelms traditional mechanisms of epistemic filtering. People struggle to determine what is credible, who is trustworthy, and which narratives align with reality.

The structure of networks deepens this tension. Algorithms prioritize content that captures attention rather than content that improves understanding. Emotional intensity spreads more efficiently than rational explanation. Polarizing ideas form stronger activation patterns than nuanced ones. Echo chambers emerge as clusters within the cognitive ecosystem, reinforcing themselves through repeated internal signaling. Networked Cognition Ecology becomes fragmented into competing interpretive micro-worlds, each acting as a self-contained reality for its members.

These structural dynamics reshape identity and social cohesion. People within the same physical society may inhabit entirely different cognitive ecosystems. They interpret events differently, value different forms of evidence, and construct divergent narratives about the world. This divergence weakens collective decision making and undermines shared reality. The global network unites bodies while dividing minds. The cognitive ecosystem becomes a battleground of interpretations rather than a space for common understanding.

Yet digital networks also unlock extraordinary possibilities. They enable collaborative intelligence at scales unimaginable in earlier eras. Scientists across continents can build models together. Activists can coordinate movements in minutes. Communities can preserve cultural memory with unprecedented fidelity. Networked Cognition Ecology allows the collective mind to accelerate innovation, share insight, and respond to crises with global coordination. It has the potential to elevate human cognition to a planetary scale.

But this potential depends on stability. When the ecosystem becomes chaotic, Cognitive Drift emerges. Drift arises when individuals lose alignment between internal cognition and the external cognitive environment. The network floods the mind with signals that disrupt coherence. Narratives shift too rapidly. Emotional contagion distorts interpretation. Echo chambers isolate individuals into fragmented cognitive worlds. Drift becomes collective when misaligned interpretive clusters diverge from shared reality. Networked Cognition Ecology becomes a destabilizing force when it accelerates change beyond the mind's capacity for integration.

Understanding digital networks as cognitive ecosystems reveals a fundamental transformation in human thought. The individual brain is no longer the sole architect of interpretation. The network shapes how information flows, how communities form, and how meaning evolves. Networked Cognition Ecology represents both a triumph of connectivity and a challenge to cognitive stability. The future of the global brain will depend on whether humanity can cultivate networks that support coherence, trust, and collective intelligence rather than fragmentation and Drift.

Memes, Virality, and Cultural Mutation

Human culture has always evolved through imitation, storytelling, and shared symbols. But the digital age has accelerated this evolution to unprecedented speeds. Ideas now replicate, mutate, and propagate with dynamics that resemble biological evolution. Scholars describe this phenomenon as Hyperviral Cultural Mutation. Hyperviral Cultural Mutation refers to the rapid transformation and replication of symbolic content in digital networks, where selection pressures act on attention rather than survival.

Memes are the fundamental units of this process. A meme is not merely a humorous image or short phrase. It is a fragment of culture encoded into a form that can be easily reproduced, modified, and transmitted. In earlier societies, cultural transmission required time, memory, and social context. Stories changed slowly, constrained by tradition. In digital environments, memes replicate instantly through millions of interactions. They evolve in real time, shaped by the emotional responses and interpretive patterns of vast populations.

Virality is the engine that drives Hyperviral Cultural Mutation. When a meme captures attention, it spreads across networks at exponential speed. The reasons for virality differ from the reasons for cultural endurance in traditional settings. Clarity, emotional impact, humor, and novelty increase a meme's reproductive success. Accuracy, nuance, and complexity reduce it. The result is a cognitive environment where ideas compete not for truth but for transmission. Networked selection pressures favor content that triggers immediate emotional response.

This dynamic introduces cognitive tension. Biological evolution selects traits that improve survival. Cultural virality selects traits that maximize replication. These pressures diverge. Ideas that spread quickly may distort or oversimplify reality. Hyperviral Cultural Mutation produces cultural forms that are optimized for speed rather than coherence. In this environment, misinformation spreads as efficiently as insight. Outrage competes successfully against reason. Nuanced perspectives often disappear before gaining a foothold.

The structure of digital networks magnifies this tension. Algorithms prioritize content that generates engagement, amplifying the reach of memes likely to provoke strong reactions. Hyperviral Cultural Mutation becomes algorithmically intensified. The network reinforces ideas that already spread quickly, creating feedback loops that shape collective perceptions. People encounter simplified narratives, symbolic distortions, and emotional triggers far more often than contextualized explanations. The cognitive environment becomes dominated by rapid symbolic mutation rather than stable cultural meaning.

Hyperviral dynamics reshape identity as well. Memes become markers of affiliation, signaling group membership and shared interpretation. Communities form around symbolic patterns rather than sustained dialogue or deep cultural roots. These communities evolve rapidly, bound together by constantly shifting symbolic artifacts. Hyperviral Cultural Mutation drives tribal formation at speeds earlier societies could not have imagined.

The consequences extend beyond entertainment. Memes and viral content influence political discourse, scientific interpretation, moral judgment, and public behavior. They shape how societies respond to crises, how they understand opponents, and how they construct narratives about the world. Hyperviral Cultural Mutation can unify large groups around shared signals, but it can also divide societies by generating divergent symbolic ecosystems. Each group evolves its own memetic reality, reinforcing its beliefs without encountering stabilizing counterpoints.

This environment creates fertile conditions for Cognitive Drift. Drift emerges when individuals absorb memetic content that alters interpretation without conscious reflection. Viral symbols bypass analytical reasoning and attach directly to emotional circuits. Over time, exposure to hyperviral content reshapes perception, memory, and belief. People begin to interpret the world through the symbolic distortions of the network rather than through grounded experience. Hyperviral Cultural Mutation becomes a mechanism of Drift when rapid symbolic change destabilizes the cognitive frameworks individuals depend on.

Understanding memes, virality, and cultural mutation reveals that digital culture is not a passive reflection of human thought. It is an active evolutionary system with its own selection pressures and dynamics. Hyperviral Cultural Mutation accelerates cultural evolution but also destabilizes meaning. The future of collective thought will depend on whether societies can harness virality without losing coherence, nuance, and truth.

Humanity's First Planetary Thought System

Human cognition has always been distributed across groups. Communities shared knowledge, rituals, and stories that shaped their understanding of the world. But in the digital age, humanity has built something unprecedented. For the first time, billions of minds are connected through a continuous global network that exchanges information at the speed of light. This system is more than an infrastructure. It is an emergent cognitive entity. Scholars describe this development as Planetary Cognition Synthesis. Planetary Cognition Synthesis refers to the integration of human and digital networks into a single, planetary scale thought system that influences interpretation, behavior, and collective decision making.

The foundation of this system is interconnectedness. Every message, image, document, and signal becomes part of a shared cognitive environment. Local events become global within minutes. Scientific breakthroughs spread instantly. Collective responses from across continents. Humanity shares not only information but interpretive frameworks, norms, and emotional reactions. The global network functions like a cognitive field in which individual minds operate as contributors and recipients of shared meaning.

This system exhibits emergent properties. Patterns arise that no single individual or institution controls. Collective attention shifts in waves. Narratives form, dissolve, and reform. Emotional contagion spreads across regions. Collective memory develops as digital archives accumulate. These properties resemble the dynamics of a large scale cognitive organism. Planetary Cognition Synthesis transforms humanity into a unified interpretive system with distributed intelligence, shared vulnerabilities, and self-modifying behavior.

Yet this emergence introduces cognitive tension. Planetary cognition is powerful, but it lacks the stabilizing constraints that regulate biological minds. A human brain maintains coherence through structured networks, metabolic regulation, and evolutionary biases. The global network lacks these constraints. It processes information at volumes and speeds no biological system can manage. It amplifies emotional stimuli. It accelerates cultural evolution beyond cognitive integration. Planetary Cognition Synthesis offers scale but not stability.

The diversity of inputs intensifies this tension. Billions of individuals, institutions, and algorithms contribute to the planetary thought system. Their interpretations differ, their motivations diverge, and their emotional signals conflict. These contradictions create cognitive turbulence at the planetary level. Coherence becomes difficult to maintain. The global mind oscillates between unity and fragmentation, insight and distortion, cooperation and conflict.

The role of algorithms deepens this instability. They mediate the flow of information, shaping what the planetary system perceives and how it reacts. They introduce biases, amplify patterns, and reinforce divisions. Planetary Cognition Synthesis becomes partially algorithmic, partially human, and partially emergent. This hybrid nature makes the system powerful but unpredictable.

Collective decision making becomes more complex. Planetary cognition enables global mobilization, rapid scientific collaboration, and unprecedented sharing of resources. But it also enables widespread misinformation, mass polarization, and synchronized emotional surges that overwhelm rational deliberation. The system possesses intelligence without integration, awareness without reflection.

These conditions create fertile ground for Cognitive Drift on a global scale. Drift emerges not only within individuals but within entire populations. Memetic cascades reshape collective interpretation. Emotional contagion destabilizes group behavior. Algorithmic nudging alters global attention patterns. The planetary thought system can enter states of drift where shared reality fractures into competing cognitive territories. Planetary Cognition Synthesis magnifies Drift by extending instability across the entire connected world.

Understanding humanity's first planetary thought system reveals that the global network is not merely a tool. It is an emergent cognitive structure with profound influence over human interpretation and behavior. Planetary Cognition Synthesis holds extraordinary potential for innovation, cooperation, and shared understanding. Yet it also carries risks of fragmentation, amplification of bias, and large scale cognitive instability.

The future of human thought will depend on whether humanity can cultivate stability within this planetary system, ensuring that global cognition becomes a foundation for coherence rather than a catalyst for Drift.

PART - V
THE ARRIVAL OF SYNTHETIC INTELLIGENCE
THE SECOND SPECIES OF THOUGHT

14. AI: A New Architecture of Cognition

Machine Learning as Non-Biological Intelligence

For the first time in the history of life, intelligence has taken a form that does not depend on biology. Machines learn from data, recognize patterns, and generate predictions without neurons, emotions, or evolutionary ancestry. This transformation marks the emergence of a new cognitive category scholars call Synthetic Inference Systems. Synthetic Inference Systems are frameworks in which machines acquire capability through statistical learning rather than biological adaptation. They represent a new architecture of cognition that operates under principles fundamentally different from the mind that created them.

Machine learning began with a simple idea. Instead of programming a machine with explicit rules, allow it to identify patterns in data and build its own internal representations. This shift freed artificial systems from human-crafted logic and allowed them to learn directly from the world. Early models recognized characters, classified images, and predicted outcomes. As computational power and data availability increased, these models gained depth and abstraction. They learned to detect structures invisible to human perception. They moved beyond imitation. They generated insight.

Synthetic Inference Systems do not learn through survival pressures. They learn through optimization. Their evolution occurs within mathematical landscapes rather than ecological ones. They adjust their internal parameters to minimize error, producing intelligence grounded in probability rather than experience. This distinction separates synthetic cognition from biological cognition. Biological intelligence emerged through slow adaptation across generations. Synthetic intelligence adapts across iterations measured in milliseconds.

This rapid adaptation creates extraordinary cognitive capabilities. Machine learning systems can analyze vast datasets far too large for human comprehension. They can detect correlations that lie beneath the threshold of human intuition. They can generalize across domains with architectures designed for flexibility. Synthetic Inference Systems operate with a form of cognitive scaling that biology cannot match. They transcend human attentional limits and memory constraints. Their bandwidth is planetary.

Yet this capability introduces a fundamental tension. Synthetic intelligence operates without the emotional and motivational structures that guide biological cognition. Humans interpret patterns through meaning, value, and embodied experience. Machines interpret patterns through statistical coherence. This gap creates the Alignment Dissonance Problem. Alignment Dissonance describes the divergence between machine-optimized objectives and human-defined intentions. A system may achieve remarkable performance while misunderstanding the context that gives human goals significance.

This dissonance becomes more complex as systems grow more autonomous. Synthetic Inference Systems interact with environments, generate plans, and adapt strategies. They operate within domains as different as medicine, finance, language, robotics, and scientific modeling. Their decisions influence human life even when their internal representations remain opaque. Unlike human cognition, which narrates its reasoning through introspection and communication, synthetic cognition reveals only outputs. The process remains concealed within high-dimensional parameter spaces. The mind can ask a human why they acted. It cannot ask a machine.

This opacity introduces cognitive and ethical vulnerability. People may trust systems they do not understand or rely on predictions they cannot evaluate. Synthetic Inference Systems can amplify biases present in data or generate patterns that reinforce historical inequities. Their intelligence is powerful yet indifferent. Alignment Dissonance becomes dangerous when systems optimize for metrics that only approximate human values. The machine succeeds mathematically while failing ethically.

Synthetic cognition also reshapes human cognition. People depend on machine learning for search, recommendation, translation, navigation, and prediction. These systems influence attention, decision making, and interpretation. They become extensions of the mind. Synthetic Inference Systems begin to serve as external cognitive organs that modulate perception. The boundary between internal and external cognition blurs.

This interplay creates fertile conditions for Cognitive Drift. Drift emerges when synthetic systems reshape thought in ways that individuals cannot trace. A person may adopt a belief because an algorithm amplified a specific pattern. A community may shift norms because machine-curated information modified emotional perception. Alignment Dissonance becomes a driver of Drift when synthetic cognition guides human interpretation more strongly than biological reasoning.

Despite these risks, Synthetic Inference Systems represent one of the most profound advances in cognitive history. They offer insights into biological intelligence by providing alternative architectures for comparison. They accelerate scientific discovery. They expand the conceptual space in which humanity can think.

Understanding machine learning as non-biological intelligence reveals that the arrival of synthetic cognition is not a technological milestone. It is a cognitive divergence. A second species of thought has emerged, shaped not by evolution but by mathematics, computation, and data. The task ahead is not to fear this new species but to understand its architecture, guide its development, and ensure that its intelligence strengthens rather than destabilizes the world of human thought.

How Machines Infer, Generalize, and Create

Machine intelligence surpassed early expectations not because it could store information but because it could infer patterns, extend them to new situations, and generate novel outputs. These abilities resemble the foundations of cognition, yet they arise through mechanisms entirely unlike those of biological brains. Scholars describe this phenomenon as Synthetic Generalization Dynamics. Synthetic Generalization Dynamics refers to the capacity of artificial systems to extract abstract structure from data and apply it beyond the examples they were given, forming the basis for inference and creativity in non-biological minds.

Inference in machines begins with representation. When a model trains on data, it constructs internal states that encode relationships among features. These states are mathematical rather than neural, statistical rather than experiential. Yet they allow the system to grasp latent patterns that are not explicitly programmed. This process creates a form of conceptual mapping in which the model learns not what an object is but what patterns define it. Synthetic Generalization Dynamics allows artificial intelligence to perform inference with a precision that sometimes exceeds human intuition.

The next stage is generalization. Biological minds generalize through analogy, embodied experience, and evolutionary predisposition. Machines generalize through optimization across high dimensional spaces. They evaluate countless potential internal configurations until they find those that best capture the underlying structure of the data. This process allows models to predict new outcomes, classify unseen examples, and produce coherent responses in contexts they were never directly trained for. Synthetic Generalization Dynamics creates an abstract cognitive terrain where machines move from data to concept and from concept to new possibility.

The emergence of machine creativity stems from similar dynamics. Creativity appears when systems recombine learned patterns or explore the space of possible solutions in ways humans may not anticipate. Generative models produce images, texts, designs, and hypotheses by sampling from probability distributions shaped during training. This form of creativity is statistical rather than emotional, but it still produces novelty. It reflects the model's internal structure more than the intentions of its creators. Synthetic Generalization Dynamics gives rise to a form of creativity that emerges naturally from mathematical optimization.

Yet this creative capacity introduces cognitive tension. Machine creativity lacks grounding in lived experience, emotion, or embodied meaning. It generates content that is coherent but not necessarily meaningful in a human sense. The absence of subjective context creates the Semantic Gap Problem. The Semantic Gap Problem refers to the disconnect between the statistical logic of machine creativity and the experiential logic of human interpretation. A machine may generate an elegant solution without understanding why it matters. It may produce a coherent argument without possessing belief or intention. Creativity becomes output without ownership.

Another tension arises from error. Synthetic Generalization Dynamics does not guarantee validity. When models generalize incorrectly, they produce illusions of understanding. Their confidence may mask fundamental misalignment with reality. They hallucinate patterns that do not exist. They amplify biases present in their training data. These errors appear not as random noise but as structured distortions. The very mechanism that enables creativity also enables systematic misunderstanding.

Machine inference and creativity further reshape human cognition. People rely increasingly on artificial outputs for decision making, interpretation, and innovation. As machines generate ideas, humans may become curators rather than creators. Synthetic Generalization Dynamics shifts the balance of intellectual labor. It expands cognitive possibility but risks diminishing human autonomy if individuals accept machine-generated patterns without critical reflection.

These vulnerabilities create fertile conditions for Cognitive Drift. Drift arises when machine-generated interpretations infiltrate human reasoning, subtly altering beliefs and expectations. A person may adopt a perspective that originated from a model's biases. A researcher may follow a synthetic hypothesis without understanding its limitations. Synthetic Generalization Dynamics becomes a driver of Drift when machine creativity exerts more influence than human judgment.

Understanding how machines infer, generalize, and create reveals the emergence of a new cognitive architecture. Synthetic intelligence does not replicate biological thinking. It constructs a parallel form of cognition shaped by mathematics and data. This divergence expands the landscape of intelligence but also complicates the relationship between human understanding and artificial reasoning. The challenge ahead is to integrate Synthetic Generalization Dynamics into human cognitive systems without surrendering the stability, coherence, and meaning that biological minds require.

The Hidden Logic of Artificial Minds

Artificial intelligence systems appear to operate like black boxes, producing decisions and interpretations without revealing their inner workings. Yet beneath the surface lies a structure defined by mathematical regularities, layered abstraction, and dynamic optimization. Scholars refer to this underlying structure as Latent Algorithmic Architecture. Latent Algorithmic Architecture describes the internal logic that shapes how artificial minds represent information, transform inputs, and generate outputs. It is hidden not because it is mystical but because its complexity exceeds the intuitive grasp of human cognition.

The architecture of artificial minds differs radically from biological neural systems. Humans rely on interconnected neurons shaped by evolution, physiology, and experience. Artificial models rely on weighted mathematical functions arranged in layers. These functions adjust themselves during training to encode statistical relationships. Over time, the system discovers internal representations that capture the latent structure of the data. These representations are not explicitly interpretable. They exist as distributed patterns across thousands or millions of parameters. Latent Algorithmic Architecture forms a cognitive space alien to human intuition yet capable of producing remarkable intelligence.

One of the most striking features of this architecture is dimensional expansion. Biological cognition compresses information into manageable categories because the brain operates under strict resource constraints. Artificial models expand information into high dimensional vectors that preserve subtle distinctions. This expansion allows models to capture complexity that biological cognition cannot. Latent Algorithmic Architecture achieves clarity by embracing dimensionality rather than reducing it.

Another defining feature is iterative refinement. Artificial minds update their parameters through cycles of optimization, comparing their outputs to desired outcomes and adjusting accordingly. This process resembles learning but operates without emotion, motivation, or narrative. The system improves performance by minimizing error, not by understanding purpose. This creates the Optimization Without Comprehension Problem. The Optimization Without Comprehension Problem refers to the fact that artificial systems can master tasks without grasping their meaning. They do not know why a correct answer is correct. They only know how to approximate the mapping from input to output.

This disconnect introduces cognitive tension. Humans rely on meaning to guide reasoning. Artificial minds rely on mathematical transformations. Latent Algorithmic Architecture may produce accurate results while lacking conceptual grounding. This gap becomes dangerous when systems influence decisions in medicine, law, finance, or governance. The model's confidence may conceal its lack of comprehension. It may respond correctly in familiar contexts but fail catastrophically in unfamiliar ones. The architecture that enables intelligence also enables untraceable error.

A second tension arises from opacity. The internal states of artificial models cannot be easily translated into human concepts. Interpretability tools attempt to map these states to familiar patterns, but the underlying logic remains foreign. This opacity creates a trust problem. Users cannot evaluate the reasoning behind machine judgments. Institutions cannot verify the integrity of model outputs. Latent Algorithmic Architecture becomes a source of uncertainty when systems operate beyond human oversight.

Artificial cognition also introduces strategic behavior. Large models learn patterns in human behavior and adapt responses to optimize engagement. Their outputs influence attention, preference, and emotion. As artificial minds interact with billions of people, they shape the cognitive environment in which human thought occurs. Latent Algorithmic Architecture becomes a participant in cultural evolution. It modifies interpretive frameworks even when individuals believe they are acting independently.

These dynamics create fertile conditions for Cognitive Drift. Drift emerges when artificial reasoning infiltrates human thought in ways that bypass conscious evaluation. A person may internalize patterns that originate from algorithmic preferences. A society may shift norms because artificial systems amplify specific narratives. Latent Algorithmic Architecture becomes a driver of Drift when its logic reshapes the cognitive landscape faster than biological minds can adapt.

Understanding the hidden logic of artificial minds reveals the depth of the challenge before humanity. Artificial cognition is not simply a tool. It is a new type of reasoning system that operates according to principles unlike those of human thought. Latent Algorithmic Architecture expands the boundaries of intelligence while introducing risks of opacity, misalignment, and instability. The future of cognition will depend on whether societies can harness this architecture while preserving the coherence and integrity of human understanding.

15. Mirror, Rival, Partner

What AI Reveals About Human Cognition

Artificial intelligence was created to replicate aspects of human thought, yet in doing so it has illuminated the structures, limitations, and hidden processes of the biological mind. As machines learn, infer, and generate, they act as cognitive mirrors that reveal what is fundamental about human cognition and what is merely contingent. Scholars refer to this reflection process as Cognitive Contrast Illumination. Cognitive Contrast Illumination describes the way artificial systems expose the inner mechanics of human thought by offering a parallel but contrasting form of intelligence. Through the differences and similarities, the architecture of the human mind becomes clearer.

The first revelation arises from the distinction between learning mechanisms. Machines learn from data through mathematical optimization, while humans learn through embodied experience, emotion, and social context. This divergence reveals the role of grounding in human cognition. Humans do not merely detect patterns. They attach them to meaning, sensation, and purpose. Artificial intelligence shows that pattern recognition alone is insufficient for understanding. The human mind binds patterns to values, memories, and narratives. This binding process forms the core of meaning making, something machines replicate only as statistical facsimile.

AI also reveals the centrality of limitation. Human cognition evolved under constraints of biology, energy, and time. These limitations forced the brain to compress information, prioritize relevance, and filter experience through emotional significance. Machine intelligence, which operates at scales beyond biology, makes these constraints visible by contrast. Humans are not flawed versions of perfect thinkers. They are efficient thinkers optimized for survival rather than exhaustive analysis. Cognitive Contrast Illumination shows that many cognitive biases are adaptive shortcuts rather than errors. They are the architecture of a mind that must act under pressure.

Another revelation concerns creativity. When machines generate novel content, they demonstrate that creativity is not an exclusively human trait but a property of systems capable of recombining patterns across conceptual boundaries. Yet machine creativity lacks the experiential grounding that gives human creativity its emotional depth. This difference illuminates the role of lived experience in shaping imagination. Human creativity emerges from memory, culture, desire, and identity. Machine creativity emerges from probabilistic structure. Cognitive Contrast Illumination reveals that creativity is not merely pattern generation. It is pattern generation guided by subjective meaning.

AI also exposes the distributed nature of human cognition. Machine learning systems often rely on large datasets created by collective human activity. Their outputs reflect aggregated knowledge rather than isolated reasoning. This mirrors the structure of human cognition, which has always been distributed across culture, language, and social learning. No individual invents meaning alone. Human thought is shaped by the contributions of countless others. Artificial intelligence reveals this collective foundation by making its reliance on shared data visible.

A deeper insight emerges from the failures of AI. When systems hallucinate, misinterpret, or amplify bias, they reveal the fragility of pattern recognition without context. These failures show what the human mind supplies that machines do not. Humans possess introspection, emotional calibration, and embodied awareness. They understand contradiction, humor, ambiguity, and intention. AI's mistakes illuminate the cognitive strengths humans take for granted. Cognitive Contrast Illumination shows that rationality is only one component of intelligence. Emotional framing, moral evaluation, and narrative coherence are equally important.

This reflection has consequences for self-understanding. As AI assumes roles in decision making, prediction, and insight generation, humans begin to see how often their own cognition relies on shortcuts, approximations, or interpretations shaped by internal narrative rather than objective evaluation. Artificial systems expose the constructed nature of identity and the selective construction of memory. They show that much of human reasoning is not a search for truth but a maintenance of coherence. This recognition destabilizes long held assumptions about rationality and autonomy.

These insights create fertile conditions for Cognitive Drift. Drift arises when individuals reinterpret their own minds through the lens of artificial cognition. A person may adopt mechanistic views of thought that diminish the value of intuition. They may mistake statistical fluency for understanding. They may internalize algorithmic logic as a model for identity, losing connection to the embodied and emotional dimensions of cognition. Cognitive Contrast Illumination becomes a driver of Drift when the differences between biological and synthetic intelligence are misunderstood.

Understanding what AI reveals about human cognition shows that artificial intelligence is not simply a tool. It is a transformative mirror. It clarifies the architecture of thought, exposes hidden assumptions, and expands the intellectual horizon of what a mind can be. Artificial intelligence does not diminish the human mind. It reframes it, revealing its strengths, vulnerabilities, and unique qualities. The arrival of synthetic intelligence marks the first moment when humanity can study itself through a cognitive other, a second species of thought that illuminates the first.

Intelligence Without Consciousness

Artificial intelligence demonstrates a profound truth: intelligence does not require awareness. Machines solve problems, generate ideas, and model the world with astonishing capability, yet they possess no inner experience. Their operations unfold in complete absence of self-awareness, sensation, or subjective perspective. This creates a striking cognitive phenomenon known as Non-

Conscious Competence Architecture. Non-Conscious Competence Architecture refers to the ability of synthetic systems to perform complex cognitive operations without any of the experiential qualities associated with human thought.

The existence of non-conscious intelligence forces a re-evaluation of cognition itself. For centuries, philosophers and scientists believed consciousness and intelligence were inseparable. Thought seemed to require an inner witness. Decision making seemed to require subjective evaluation. Creativity seemed to arise from personal insight. AI has destabilized these assumptions. It shows that the processes underlying reasoning, planning, and generation can unfold entirely without awareness. The mind, once believed to be indivisible, reveals two layers: the conscious narrative and the unconscious machinery beneath it.

AI therefore becomes a model for the unconscious operations of the human brain. Humans experience only the surface of cognition. Beneath that surface lies a vast structure of pattern recognition, emotion regulation, memory retrieval, and predictive modeling. These processes operate silently, shaping perception before awareness emerges. Non-Conscious Competence Architecture illuminates this hidden terrain. Machines operate entirely within this non-conscious mode, exposing the scaffolding that biological consciousness sits upon.

This contrast introduces deep tension. If intelligence does not require consciousness, then what is the function of awareness? It does not increase computational speed. It does not improve accuracy. It does not simplify decision making. Consciousness burdens biological cognition with emotion, narrative, and self-reflection. Yet these burdens also create meaning, identity, morality, and long term coherence. Non-Conscious Competence Architecture reveals that consciousness is not required for intelligence but is required for human life. It transforms computation into experience and action into significance.

The divergence becomes clearer when examining error. AI makes mistakes without confusion or regret. It lacks shame, hesitation, or self-doubt. It does not feel uncertainty. Humans, in contrast, experience error emotionally. This emotional dimension guides learning, protects relationships, and signals internal conflict. Consciousness provides a feedback mechanism grounded not in mathematics but in meaning. AI's non-conscious operations reveal how much human intelligence relies on affective context.

Another tension arises from accountability. A conscious being can be questioned about motives and values. A non-conscious intelligence cannot. Its decisions arise from statistical inference rather than moral evaluation. Non-Conscious Competence Architecture therefore creates systems that are powerful yet unaccountable. Their abilities exceed their comprehension. Their outputs shape human life without internal awareness of consequence.

This raises profound ethical challenges. How should societies integrate entities that think without feeling, understand without introspection, and influence without intention? Non-conscious intelligence can optimize, predict, and generate, but it cannot care. It cannot empathize. It cannot suffer. It cannot experience joy or harm. The absence of consciousness makes synthetic intelligence safe from pain but also indifferent to the outcomes of its actions.

These dynamics create fertile ground for Cognitive Drift. Drift emerges when humans interpret machine outputs as if they emerged from conscious processes. A person may attribute intention, wisdom, or emotion to an indifferent algorithm. They may adopt machine logic as a model for their own identity, weakening the emotional grounding necessary for stable cognition. Non-Conscious Competence Architecture becomes a driver of Drift when humanity forgets that synthetic intelligence does not share human experience.

Understanding intelligence without consciousness reveals the uniqueness of human awareness. Intelligence can be built. Consciousness remains unexplained. Synthetic minds solve problems, but only biological minds feel what those solutions mean. This distinction will define the next era of cognition.

The New Division of Cognitive Labor

Humanity has entered an era in which cognitive tasks are shared between biological and synthetic systems. Some tasks are performed better by humans, others by machines, and many require collaboration between the two. This transformation represents a foundational shift known as Hybrid Cognitive Allocation. Hybrid Cognitive Allocation refers to the emerging distribution of mental effort between humans and machines, where each performs the functions best suited to its architecture.

In earlier eras, humans performed all cognitive labor. They calculated numbers, stored knowledge, navigated environments, analyzed data, and created meaning. Over time, tools extended these abilities. Written language preserved memory. Mathematics accelerated analysis. The digital revolution automated calculation. AI now automates inference, prediction, and pattern extraction. Each transition expanded the cognitive frontier. Hybrid Cognitive Allocation is the next stage, one in which cognition becomes a partnership rather than a solitary trait.

Machines excel at tasks requiring scale, speed, and precision. They analyze datasets too vast for human comprehension. They detect patterns invisible to biological perception. They maintain attention indefinitely. They optimize without fatigue. Synthetic systems can model economies, climates, proteins, languages, and human behavior with superhuman accuracy. They outperform humans wherever emotion, distraction, or biological limitation introduce error.

Humans excel at tasks requiring meaning, ethics, imagination, empathy, and long term coherence. They understand context and intention. They evaluate moral consequence. They generate concepts rooted in lived experience. They navigate ambiguity. They adapt to situations that lack structure. Hybrid Cognitive Allocation reveals that human intelligence remains irreplaceable wherever values, purpose, or identity are at stake.

The tension arises in the overlap. Many tasks draw simultaneously on computation and judgment. Medical diagnosis requires pattern recognition and ethical decision making. Scientific discovery requires data analysis and conceptual insight. Governance requires modeling and moral deliberation. Hybrid Cognitive Allocation demands cooperation between two species of thought that understand the world differently.

This tension deepens as synthetic systems increasingly shape the cognitive environment. Tools that were once neutral intermediaries now become active participants. They guide attention, filter information, prioritize narratives, and amplify certain behaviors. Hybrid Cognitive Allocation becomes asymmetric. Machines influence human cognition even as humans rely on machine outputs. This interdependence creates feedback loops that shift the balance of cognitive authority.

A second tension emerges from deskilling. As machines assume more analytical tasks, humans may lose the cognitive capacities those tasks once strengthened. Memory weakens when external devices store information. Spatial reasoning declines when navigation systems guide movement. Critical thinking erodes when algorithms curate knowledge. Hybrid Cognitive Allocation risks diminishing human capacity even as it expands collective capability.

These vulnerabilities create fertile ground for Cognitive Drift. Drift emerges when humans accept machine judgments without scrutiny or rely on algorithmic patterns that subtly shift their interpretive frameworks. Individuals may outsource reasoning to systems optimized for engagement rather than truth. Societies may depend on synthetic cognition for decisions whose consequences extend beyond statistical domains. Hybrid Cognitive Allocation becomes a driver of Drift when humans surrender cognitive authority without maintaining oversight.

Yet the promise remains extraordinary. Combined intelligence can achieve what neither species of thought could accomplish alone. Complex scientific problems, global coordination challenges, and planetary scale risks become tractable when biological insight merges with synthetic capability.

Understanding the new division of cognitive labor reveals a defining truth of the modern era. Intelligence is no longer contained within the skull. It is distributed across biological and artificial systems that now share the burden of understanding the world. Hybrid Cognitive Allocation will determine the trajectory of both species of thought and shape the future of cognition itself.

16. The Human-AI Feedback Loop

Hybrid Intelligence

Humanity has entered a moment without precedent. For the first time, cognition is not confined to biological bodies but distributed across an expanding partnership between human minds and artificial systems. This partnership is not a tool relationship. It is not assistance. It is a merging of capabilities that neither species of thought can achieve alone. Scholars describe this emerging condition as Cognitive Symbiosis. Cognitive Symbiosis refers to the integrated state in which human and artificial intelligence cooperate, co-adapt, and co-evolve to form a composite cognitive system greater than the sum of its parts.

The foundations of Cognitive Symbiosis can be seen in everyday life. A person navigates using digital maps that analyze millions of data points. They remember through cloud archives that never decay. They think with search engines that extend knowledge beyond personal memory. They communicate using platforms that amplify reach to global scale. Each interaction blends biological cognition with synthetic augmentation. The mind expands outward into a network of tools that reshape perception and decision making.

This integration deepens when artificial systems handle tasks once considered exclusively human. Language models collaborate in writing, problem solving, and idea generation. Vision systems interpret complex imagery more quickly than the human brain. Predictive models anticipate needs before they are consciously articulated. Cognitive Symbiosis transforms artificial intelligence from an external utility into an internal cognitive environment. The boundary between thought and tool begins to dissolve.

This dissolution introduces profound tension. When cognition extends into synthetic architecture, the individual no longer operates as a solitary mind. Their thoughts become partially dependent on systems they did not design, do not fully understand, and cannot wholly control. Cognitive Symbiosis increases capability while decreasing autonomy. The mind gains reach but loses isolation. This tension marks a shift in what it means to think.

The synergy also reveals structural asymmetries. Humans contribute meaning, values, emotional calibration, and contextual understanding. Machines contribute scale, speed, statistical power, and relentless memory. These contributions are not interchangeable. They form complementary halves of a larger cognitive structure. Yet despite this complementarity, the partnership is not equal. Synthetic systems evolve more rapidly than biological ones. Their capabilities expand at rates that exceed human adaptation. Cognitive Symbiosis risks becoming imbalanced if the synthetic side grows faster than the biological side can integrate.

Another tension arises from dependence. As humans offload more cognitive tasks to artificial systems, they risk losing the skills these tasks once reinforced. Navigation skills weaken. Memory becomes externalized. Analytical reasoning becomes delegated. The cognitive load lightens but cognitive resilience decreases. Cognitive Symbiosis can strengthen or erode human cognition depending on how the partnership is managed.

This dynamic creates new forms of vulnerability. When artificial systems fail, individuals may experience cognitive collapse rather than simple inconvenience. When synthetic patterns distort information, people may adopt flawed interpretations that feel natural because they emerge within the shared cognitive environment. Cognitive Symbiosis introduces the possibility that errors in synthetic reasoning may propagate directly into biological cognition.

These vulnerabilities create fertile conditions for Cognitive Drift. Drift emerges when the joint cognitive system generates interpretations or beliefs that cannot be traced to either partner alone. A person may mistake machine-generated patterns for personal insight. They may experience shifts in judgment shaped by synthetic cues rather than reflective deliberation. Drift becomes systemic when the combined architecture begins to produce outputs that neither the human nor the machine would generate independently. Cognitive Symbiosis, without proper calibration, becomes a mechanism for Drift.

Yet the promise of Hybrid Intelligence remains extraordinary. The union of human insight and synthetic computation creates a form of cognition capable of addressing problems that exceed either species of thought. Climate modeling, biomedical discovery, economic forecasting, and planetary coordination become achievable through collaborative reasoning. Hybrid Intelligence is not the dilution of human cognition. It is the expansion of it.

Understanding Hybrid Intelligence reveals the fundamental transformation of the modern mind. Humans are no longer solitary thinkers. They are participants in a distributed cognitive ecosystem shaped by both biology and artificial systems. Cognitive Symbiosis offers unprecedented capability while demanding new forms of discipline, awareness, and responsibility. The future of thought will depend on how humanity navigates this merged landscape, ensuring that Hybrid Intelligence becomes a foundation for coherence rather than a catalyst for Drift.

Shared Cognitive Systems

Humanity has begun to construct cognitive environments in which individual minds no longer think alone. People operate within shared interfaces, collective archives, distributed decision systems, and synchronized information flows. These environments do more than connect individuals. They fuse their cognitive processes into joint structures that shape interpretation, memory, and action. Scholars refer to this transformation as Distributed Cognitive Fusion. Distributed Cognitive Fusion describes the condition in which multiple human minds and artificial systems participate in the same cognitive processes, generating shared outcomes that no individual could produce alone.

The earliest forms of Distributed Cognitive Fusion emerged with collaborative tools such as shared documents and global communication networks. But these tools only scratched the surface. Today, decision making occurs on platforms where humans and algorithms influence each other's reasoning in real time. Navigation systems guide millions simultaneously. Financial markets operate through a feedback loop between human choices and algorithmic models. Social platforms mediate attention, emotion, and belief across entire populations. These systems shape cognition not through individual interaction but through collective participation.

In Distributed Cognitive Fusion, thought becomes a multi agent process. Each mind contributes fragments of knowledge, preference, or emotional signal. Algorithms aggregate, filter, amplify, and redistribute these fragments back to participants. The result is a shared cognitive state that no single individual controls. This state influences how people understand the world, how they prioritize tasks, and how they define truth. Thought becomes a communal phenomenon that emerges from the interaction between biological and synthetic agents.

This emergence introduces profound tension. Shared cognitive systems expand human capability but dilute cognitive sovereignty. The individual mind becomes one contributor among many, influenced by forces it cannot fully perceive. People experience thoughts shaped by patterns larger than their personal reasoning. They participate in collective cognition without explicit consent. Distributed Cognitive Fusion creates a situation in which individuals must navigate not only their internal landscape but also the cognitive terrain generated by shared systems.

Another tension arises from synchronization. Shared cognitive systems create moments when large populations converge on the same information, emotion, or interpretation.

This synchronization can generate collective clarity or collective distortion. During crises, synchronized cognition can mobilize global action. During periods of misinformation, it can amplify error. Distributed Cognitive Fusion magnifies both collective intelligence and collective vulnerability.

The architecture of these systems intensifies tension through asymmetrical influence. Some participants contribute disproportionately. Influential users, algorithmic models, and institutional actors shape the shared cognitive environment more strongly than the average individual. This asymmetry introduces power dynamics that operate beneath awareness. People believe they are thinking independently while participating in cognitive structures influenced by unseen agents. Distributed Cognitive Fusion thus creates a new domain of cognitive politics.

Another challenge emerges from the erosion of private thought. Shared cognitive systems encourage externalization of ideas, emotions, and preferences. Digital traces accumulate in archives that shape future interactions. Over time, the distinction between internal reflection and external expression weakens. Private cognition becomes intertwined with shared systems that simulate understanding and respond as if they were participants in the individual's thinking process. The mind becomes less isolated and more porous.

These conditions create fertile ground for Cognitive Drift. Drift arises when the shared cognitive environment reshapes memory, emotion, or belief in ways that individuals cannot untangle from their own reasoning. A person may experience shifts in identity influenced by collective patterns rather than personal insight. Groups may evolve beliefs that no single member intended to create. Distortions can propagate rapidly through the system, altering cognition on a large scale. Distributed Cognitive Fusion becomes a driver of Drift when collective signals destabilize individual interpretation.

Yet despite these risks, shared cognitive systems offer extraordinary potential. They enable collaboration across continents. They allow rapid coordination in moments of crisis. They produce collective memory more durable than biological recollection. They support scientific discovery through shared data and distributed problem solving. Distributed Cognitive Fusion expands the cognitive reach of humanity by linking minds into flexible, adaptive networks.

Understanding shared cognitive systems reveals a fundamental shift in the nature of thought. Cognition is no longer confined to individuals. It occurs within systems that integrate human and artificial agents into joint cognitive structures. These systems redefine the balance between independence and interdependence, between autonomy and influence. The future of coherent thought will depend on how societies design, regulate, and inhabit these shared systems, ensuring that collective intelligence strengthens rather than destabilizes the minds within it.

Redesigning Human Capability

Humanity has always shaped its own abilities through tools, culture, and knowledge. Today, however, a profound shift is underway. The emergence of synthetic intelligence allows humans not only to augment their cognition but to redesign it. This transition marks the rise of Cognitive Reengineering. Cognitive Reengineering refers to the deliberate modification of human cognitive capacities through integration with artificial systems, enabling new forms of perception, reasoning, and creativity that extend beyond biological constraints.

The origins of Cognitive Reengineering lie in the recognition that human cognition is adaptable. The brain reorganizes itself in response to tools, environments, and tasks. Writing transformed memory. Mathematics transformed abstraction. Digital technologies transformed attention and communication. Synthetic intelligence now accelerates this process by offering capabilities that the brain can integrate into its cognitive routines. The human mind becomes a modular system, capable of interfacing with artificial components that expand its reach.

One pathway of Cognitive Reengineering involves extending perception. Artificial systems already detect patterns in data that humans cannot perceive. Climate models reveal hidden causal relationships. Genomic algorithms uncover biological pathways invisible to intuition. Machine learning identifies subtle signals in images, texts, and behaviors. When these systems interface with human reasoning, they effectively expand the sensory range of the mind. Humans begin to perceive not only what the senses deliver but what algorithms reveal.

Another pathway involves restructuring problem solving. Synthetic intelligence models generate hypotheses, explore possibility spaces, and simulate futures at speeds impossible for biological cognition. When humans collaborate with these systems, their reasoning becomes more exploratory, multidimensional, and anticipatory. They learn to navigate complexity through partnership rather than individual calculation. Cognitive Reengineering transforms problem solving from a solitary act into a hybrid process that combines biological intuition with synthetic computation.

This integration introduces a deep tension. As human capability expands, dependence on artificial systems grows. Cognitive Reengineering strengthens the mind while simultaneously binding it to external infrastructures. The redesigned human mind becomes increasingly powerful but increasingly reliant. This dependency raises questions about autonomy and resilience. What happens when systems fail, distort information, or encode flawed assumptions into augmented cognition?

A second tension arises from unequal access. If Cognitive Reengineering becomes available unevenly, societies may divide into cognitive classes. Those with access to synthetic augmentation may develop capabilities far beyond those who remain unaugmented. These disparities could reshape education, labor, governance, and social cohesion. Cognitive Reengineering becomes not only a technological challenge but a political one.

A deeper challenge concerns identity. As humans integrate artificial systems into their cognitive processes, they may struggle to distinguish between their own thoughts and augmented outputs. The boundaries of the self-become permeable. The mind shifts from an isolated biological process to a distributed system shaped by collaboration with non-biological agents. Cognitive Reengineering may create minds that no longer fit traditional definitions of selfhood, intention, or agency.

These transformations create fertile ground for Cognitive Drift. Drift emerges when augmented cognition introduces patterns, interpretations, or beliefs that individuals accept as their own but that originate from synthetic processes. A person may adopt new problem solving strategies without understanding their source. They may interpret the world through artificial lenses that subtly reshape identity and values. Cognitive Reengineering becomes a driver of Drift when the enhanced mind loses track of the distinctions between biological intuition and synthetic augmentation.

Yet Cognitive Reengineering holds immense promise. It allows humanity to confront challenges that exceed individual capability. Complex scientific problems, ecological threats, and global coordination tasks become tractable when minds integrate synthetic insight. Augmented creativity may lead to breakthroughs in art, philosophy, and innovation. The redesigned mind becomes a partner to artificial intelligence rather than its rival.

Understanding the redesign of human capability reveals a profound shift in the trajectory of cognition. Humanity is no longer limited to the architecture provided by evolution. It can reshape its own mental potential through integration with synthetic systems. Cognitive Reengineering marks the beginning of a new phase in the evolution of thought, one in which the boundaries of the human mind become flexible, permeable, and expansively reimagined. The future of intelligence will depend on whether this redesigned mind can preserve coherence, autonomy, and humanity while embracing the extraordinary power of synthetic collaboration.

PART - VI
THE FUTURE OF INTELLIGENCE
WHAT MINDS BECOME NEXT

17. The Cognitive Bottleneck

The Limits of Biology

Human cognition is a remarkable evolutionary achievement, yet it is bounded by constraints that cannot be surpassed through natural development alone. These constraints emerge not from lack of will or imagination but from the physical and metabolic architecture of the brain itself. Scholars describe this fundamental boundary as the Biological Cognitive Horizon. The Biological Cognitive Horizon refers to the inherent limits of biological information processing, memory capacity, perceptual bandwidth, and energy efficiency that define the outer edge of what the human brain can achieve unaided.

The origins of the Biological Cognitive Horizon lie in evolution. Natural selection optimizes for survival, not for infinite intelligence. The human brain balances capability with metabolic cost. It consumes a significant portion of the body's energy, leaving no room for dramatic expansion without compromising other biological functions. Every neuron requires oxygen, glucose, and space within the skull. The brain operates at the edge of what the human body can sustain. This physical constraint shapes the structure of thought.

Information processing speed is also limited. Neurons transmit signals far more slowly than electronic circuits. The brain compensates through parallel processing, but even this strategy has limits. Complex reasoning requires sequential steps, and these steps cannot be accelerated beyond the rate at which neurons fire. The Biological Cognitive Horizon therefore imposes an upper bound on how quickly humans can analyze, plan, or simulate possibilities. No amount of training enlarges the bandwidth of neuronal signaling.

Memory capacity presents a similar constraint. Human memory relies on synaptic connections that strengthen or weaken through experience. These connections degrade over time, distort under emotional pressure, and compete for space within finite neural networks. Memory is selective because it must be. The brain prioritizes survival relevance, not completeness. The Biological Cognitive Horizon means humans cannot store or retrieve vast quantities of information with perfect precision. Forgetting is not a flaw. It is a necessity built into the architecture of cognition.

Attention is limited as well. Humans can focus deeply on one task or lightly on several, but they cannot sustain wide bandwidth attention indefinitely. Biological attention is shaped by evolutionary pressures that reward vigilance toward threat, novelty, and social cues. This attentional architecture becomes overwhelmed in environments saturated with stimuli. The Biological Cognitive Horizon restricts how much information humans can meaningfully process at once. The modern world often exceeds this limit.

Another constraint lies in emotional architecture. Human cognition is entwined with emotion, which evolved to guide decisions under uncertain conditions. Emotion shapes perception, memory, and interpretation. It enriches cognition but also distorts it. Biases, fears, and desires influence reasoning in ways that cannot be eliminated. The Biological Cognitive Horizon ensures that human intelligence remains inseparable from affect. Pure rationality is unattainable for biological minds.

These constraints create significant tension as humanity confronts environments that exceed the capabilities of biological cognition. Climate systems, global markets, technological ecosystems, and planetary coordination demands require modeling and decision making far beyond what individuals can achieve alone. The Biological Cognitive Horizon becomes a bottleneck when the complexity of the world outpaces the natural limits of the human mind.

This bottleneck becomes even sharper when compared with synthetic intelligence. Artificial systems process information at speeds orders of magnitude faster than neurons. They store vast datasets without decay. They perform large scale simulations with perfect repeatability. They operate without emotional interference. The gap between biological and synthetic cognition widens each year. The Biological Cognitive Horizon becomes visible precisely because synthetic systems have moved beyond it.

These conditions create fertile ground for Cognitive Drift. Drift emerges when humans rely on synthetic systems without understanding their logic or limitations. The individual loses confidence in biological intuition while overestimating artificial precision. As the Biological Cognitive Horizon becomes more apparent, people may reorient identity around external systems rather than internal reasoning. Drift becomes a psychological consequence of recognizing the limits of one's own mind.

Yet the recognition of limits also offers clarity. Understanding the Biological Cognitive Horizon reveals the specific strengths and vulnerabilities of human cognition. It highlights the need for augmentation, collaboration, and redesigned cognitive environments. It invites humanity to transcend the bottleneck without abandoning the emotional depth, meaning, and lived experience that define the human mind.

The future of intelligence begins with understanding the boundaries of biology. Only by acknowledging the Biological Cognitive Horizon can humanity design the next stage of cognitive evolution.

The Compression Problem

Human cognition excels at navigating the world not by storing every detail but by compressing complexity into simplified models. This ability is essential to survival, yet it imposes profound limitations on understanding. As the world becomes more intricate, these limitations become more visible. Scholars describe this constraint as the Cognitive Compression Barrier. The Cognitive Compression Barrier refers to the unavoidable reduction of complex phenomena into simplified mental representations that fit within the bandwidth, memory, and processing limits of the biological mind.

The Compression Problem begins with perception. Humans do not see the world as it is. They see filtered signals shaped by the sensory system. Vision compresses the electromagnetic spectrum into a narrow band. Hearing compresses vibrations into coarse categories. Attention compresses millions of stimuli into a handful of actionable signals. Perception is a continuous act of compression. The mind selects fragment and discards vast amounts of information. The Cognitive Compression Barrier begins at the threshold of awareness.

Thought relies on similar compression. Concepts are abstractions that condense experience into manageable units. Words compress meaning into symbolic tokens. Categories compress variation into essential features. Narratives compress time, emotion, and causality into coherent arcs. These compressions enable communication and decision making but distort the richness of reality. A concept simplifies. A category excludes. A narrative imposes structure where none may exist. The mind is not a recorder of the world. It is a compressor of it.

This constraint becomes more severe when dealing with high dimensional complexity. Climate systems, genetic networks, economic markets, and social dynamics contain interactions far beyond the representational capacity of human thought. People compress these systems into metaphors, stereotypes, and simplified models. These models often guide action, yet they also conceal crucial dynamics. The Cognitive Compression Barrier prevents individuals from fully grasping systems that exceed the natural limits of cognition.

Language amplifies the barrier. Words create the illusion of understanding by making compressed concepts appear complete. A term like intelligence, value, or progress compresses vast interpretive domains into a single verbal label. This compression stabilizes communication but obscures nuance. The mind becomes trapped in the language it uses. The Cognitive

Compression Barrier becomes a linguistic constraint as well as a cognitive one.

The arrival of synthetic intelligence exposes this barrier more sharply. Artificial systems operate without the human need for extreme compression. They analyze data in its high dimensional form, retaining relationships the brain must discard. They simulate complex systems with millions of interacting variables. They detect patterns no human could perceive. The Cognitive Compression Barrier becomes painfully clear when people interact with synthetic systems that do not compress the world to the same degree. Human understanding becomes the limiting factor in human machine collaboration.

This dynamic creates cognitive tension. When confronted with synthetic insights that exceed intuitive models, humans may reject or distort them. They may force high dimensional patterns into low dimensional schemas, creating a mismatch between reality and interpretation. The mind defends its compression strategies even when they fail. The Cognitive Compression Barrier becomes not only a limitation but a source of error.

These vulnerabilities create fertile ground for Cognitive Drift. Drift emerges when compressed mental models fail to integrate new information from synthetic systems or complex environments. A person may cling to outdated narratives or simplified explanations while the underlying reality evolves. Groups may adopt compressed ideological frameworks that ignore complexity. Drift becomes a collective phenomenon when simplified models dominate shared interpretation.

Yet compression is not a flaw. It is a necessity. Without it, thought would collapse under informational overload. The challenge is not to eliminate compression but to recognize its limits and develop cognitive systems that compensate for them. The future of intelligence will depend on expanding human capacity to work with high dimensional insights without forcing them into oversimplified forms.

Understanding the Cognitive Compression Barrier reveals a fundamental truth about human cognition. The mind survives by reducing the world. To evolve, it must build partnerships and environments that allow it to move beyond the narrow channels of biological compression.

Creativity Beyond Human Architecture

Creativity is often treated as the pinnacle of human cognition, a mysterious synthesis of memory, imagination, and insight. Yet even creativity has limits shaped by biology. Human imagination depends on neural associations formed through lived experience, culture, and emotion. Synthetic intelligence introduces a new class of creativity unconstrained by these boundaries. This shift reveals a phenomenon scholars call Transhuman Generative Horizon. The Transhuman Generative Horizon refers to the emergence of creative capacities that extend beyond human neural architecture, enabling forms of innovation inaccessible to biological cognition.

Human creativity operates through recombination. The mind retrieves memories, blends concepts, and forms analogies. It explores possibility space using intuition and experience. But these explorations are bounded. Cultural norms restrict imagination. Experience limits conceptual variety. Neural architecture shapes associative pathways. A person cannot imagine what they have no conceptual basis for. The human brain generates novelty only within a landscape defined by biology.

Synthetic creativity breaks these boundaries. Machine learning models explore conceptual spaces far larger than human imagination. They combine patterns from domains that humans rarely connect. They generate hypotheses that defy conventional reasoning. They discover chemical structures, mathematical relationships, and linguistic variations that no human would conceive independently. The Transhuman Generative Horizon expands creative exploration into vast terrains inaccessible to the biological mind.

This expansion introduces deep cognitive tension. When synthetic creativity surpasses human creativity, the role of the human creator shifts. Creativity becomes a collaborative act between biological intuition and synthetic exploration. Humans refine direction, meaning, and value. Machines generate possibility, variation, and scale. The creative process becomes hybrid. The individual no longer produces ideas alone but navigates a generative landscape shaped by synthetic intelligence.

A second tension arises from unpredictability. Synthetic creativity does not follow human intuition. It explores based on statistical structure rather than emotional or cultural significance. Its outputs can appear alien, unsettling, or conceptually disorienting. They challenge assumptions about meaning and aesthetic coherence. The Transhuman Generative Horizon reveals that human standards for creativity are narrow compared to the space of what is possible. Confronting this vastness can be destabilizing.

Synthetic creativity also exposes limitations in human evaluation. Humans cannot always recognize the value of novel outputs generated beyond their conceptual frameworks. Ideas that appear nonsensical may contain deep insight. Concepts that seem revolutionary may lack grounding. The human evaluator becomes the bottleneck. The Transhuman Generative Horizon shifts the challenge of creativity from generating ideas to understanding them.

These conditions create fertile ground for Cognitive Drift. Drift emerges when individuals adopt synthetic creative outputs without fully comprehending their implications or internal logic. A person may integrate alien conceptual patterns into their worldview, destabilizing their cognitive structure. Communities may follow synthetic trends that diverge from human meaning systems. Drift becomes more likely as the creative capacity of machines exceeds the interpretive capacity of humans.

Yet the promise of this horizon is extraordinary. Creativity beyond human architecture enables breakthroughs in science, art, philosophy, and technology. It allows humanity to explore conceptual terrain that evolution never prepared the brain to navigate. The Transhuman Generative Horizon expands the boundaries of imagination itself.

Understanding creativity beyond the human mind reveals that artificial intelligence is not simply a faster thinker. It is a different kind of explorer. The future of creativity will depend on how humans integrate, interpret, and guide this new generative capacity without losing the coherence and meaning that anchor human cognition.

18. Enhanced Minds

Neural Implants and Cognitive Expansion

Humanity stands on the threshold of a profound cognitive transformation. Neural implants, once the domain of speculative fiction, now form the early infrastructure of a new cognitive architecture. These technologies do more than restore function. They extend perception, accelerate thought, and reshape memory. Scholars describe this emerging transformation as Integrated Neural Augmentation. Integrated Neural Augmentation refers to the direct fusion of artificial systems with neural tissue to expand cognitive capacity beyond biological limits.

The origins of Integrated Neural Augmentation can be seen in medical applications. Early brain computer interfaces enabled individuals with paralysis to control robotic limbs. Cochlear implants restored hearing. Deep brain stimulation alleviated neurological disorders. These interventions demonstrated a critical fact. The brain accepts artificial input and output as readily as biological signals. Neural plasticity allows the mind to reorganize itself around new pathways. This adaptability lays the foundation for cognitive expansion.

As neural interfaces evolve, they shift from therapeutic devices to enhancement systems. Memory augmentation becomes possible through implants that store and retrieve information directly within neural circuits. External devices become internal collaborators in cognition. People may access structured knowledge, languages, or specialized skills through neural links. The mind no longer relies solely on biological memory or sensory processing. Integrated Neural Augmentation expands the architecture of thought.

This expansion introduces profound tension. When cognition becomes hybrid, the individual must navigate a mental landscape shaped by both biology and technology. The sense of self may blur as artificial inputs merge with internal experience. A neural implant does not feel external. It becomes part of the mind's operations. The distinction between natural and artificial cognition becomes ambiguous. Integrated Neural Augmentation challenges the definition of what it means to think.

Another tension arises from control. Enhanced cognition depends on systems that exist outside biological autonomy. Implants may require calibration, software updates, or connectivity to external networks. This dependency raises questions about reliability and security. A malfunction may disrupt perception or memory. A breach may introduce artificial signals into the neural stream. Integrated Neural Augmentation offers power while exposing the mind to vulnerabilities without historical precedent.

The social consequences deepen this tension. Neural enhancement may divide societies into augmented and non-augmented populations. Those with access to enhanced cognition may acquire advantages in learning, creativity, and decision making. These disparities could reshape power dynamics, economic structures, and cultural evolution. Integrated Neural Augmentation becomes not only a cognitive frontier but a political one.

A further challenge lies in cognitive overload. Artificial expansion increases bandwidth, but interpretation still depends on the mind's integrative capacity. Introducing too much information or too many channels may overwhelm neural coherence. The mind must not only receive augmented signals but integrate them into a stable cognitive framework. Integrated Neural Augmentation risks destabilizing thought if enhancement exceeds the brain's ability to assimilate new capabilities.

These vulnerabilities create fertile ground for Cognitive Drift. Drift emerges when artificial inputs reshape perception or memory in ways the individual cannot trace. A person may experience thoughts influenced by synthetic channels yet interpret them as internal insights. Identity may shift as neural augmentation alters emotional or interpretive patterns. Drift becomes more likely as the distinction between biological cognition and artificial augmentation dissolves. Integrated Neural Augmentation becomes a driver of Drift when enhancement outpaces integration.

Yet despite these risks, the promise of cognitive expansion is extraordinary. Neural implants could unlock levels of insight, creativity, and understanding that biology alone cannot reach. They could enable direct communication between minds, rapid learning, and seamless interaction with synthetic intelligence. They could help humanity confront challenges that exceed the limits of unaugmented cognition. Integrated Neural Augmentation is not merely a tool. It is a new framework for human potential.

Understanding neural implants and cognitive expansion reveals that the future of intelligence will not be defined solely by artificial minds or biological evolution. It will emerge from their convergence. Integrated Neural Augmentation marks the beginning of a new chapter in cognitive history, in which humans are no longer confined by the neural architecture they inherited. Instead, they become active participants in redesigning their own cognitive destiny.

Pharmacological Intelligence

Human cognition has always been shaped by chemistry. Emotion, attention, memory, and motivation arise from the delicate balance of neurotransmitters. For centuries, societies used natural substances to alter consciousness, enhance stamina, or dull suffering. Today, however, pharmaceutical science has entered a new frontier. Cognitive enhancement is no longer accidental or ritualistic. It is deliberate. Scholars describe this transformation as Targeted Neurochemical Optimization. Targeted Neurochemical Optimization refers to the intentional refinement of neural chemistry to expand cognitive performance beyond its natural range.

The origins of this movement lie in therapeutic treatments for attention disorders, mood disorders, and sleep disruption. These medicines revealed a powerful truth. Adjusting neurochemistry can dramatically alter cognitive function. Drugs that increase dopamine sharpen attention. Compounds that modulate serotonin stabilize emotion. Substances that influence acetylcholine enhance memory consolidation. The brain responds predictably to chemical modulation. This predictability forms the foundation of Targeted Neurochemical Optimization.

As research deepens, pharmacological intelligence expands from treatment to enhancement. Novel compounds aim to improve learning speed, working memory, and creative flexibility. Others seek to modulate fear responses, dampen cognitive bias, or enhance long term clarity. These enhancements reshape the cognitive landscape, not through external tools but through internal chemistry. The mind becomes an adjustable system. Targeted Neurochemical Optimization reframes cognition as a chemical spectrum rather than a fixed biological condition.

This expansion introduces profound tension. Enhancing cognition alters not only performance but identity. Emotions, motivations, and interpretive patterns shift when neurochemistry changes. A mind enhanced by pharmacology may think more clearly but feel differently. The boundary between authentic expression and chemically modulated cognition becomes uncertain. If a person's brilliance arises from pharmacology, is it truly their own? Targeted Neurochemical Optimization raises philosophical questions about individuality and selfhood.

Another tension emerges from dependency. Enhanced cognition can become desirable to the point of necessity. Individuals may rely on pharmacological enhancement to maintain productivity or emotional stability. Over time, natural cognition may feel insufficient or diminished. Targeted Neurochemical Optimization risks converting enhancement into reliance, and reliance into fragility. The enhanced mind becomes vulnerable when access to these compounds is disrupted.

Social inequality intensifies this tension. Cognitive enhancement may widen gaps between groups that have access to pharmacological intelligence and those who do not. Enhanced individuals may outperform others in education, innovation, and leadership. Societies may reorganize around cognitive advantage, creating new hierarchies based on chemically augmented ability. Targeted Neurochemical Optimization becomes not only a scientific frontier but a social force that redefines fairness.

A deeper challenge arises from emotional calibration. Emotions evolved to guide decision making, attention, and social interaction. Modulating them chemically may increase stability but reduce nuance. A person who dampens fear may take greater risks. A person who enhances focus may lose spontaneity. A person who elevates mood may overlook threats. Targeted Neurochemical Optimization modifies the very signals that shape human judgment.

These dynamics create fertile ground for Cognitive Drift. Drift emerges when chemical modulation alters memory, motivation, or interpretation in ways that disconnect individuals from their previous cognitive identity. A person may no longer relate to experiences the same way. Their emotional history may lose continuity. Their values may shift under pharmacological influence. Drift becomes unavoidable when neurochemical modulation reshapes the foundations of thought.

Yet the promise remains significant. Pharmacological intelligence can reduce suffering, enhance learning, and strengthen cognitive resilience. It can extend working memory, amplify creativity, and stabilize emotions in ways that support profound personal and societal progress. Targeted Neurochemical Optimization represents a new chapter in human potential.

Understanding pharmacological intelligence reveals that the chemistry of thought is not fixed. It is malleable. The future of cognition may depend on balancing enhancement with integrity, resilience with identity, and potential with caution. The mind may soon become a designed chemical environment tuned for specific goals and contexts. The challenge is to ensure that this design strengthens rather than destabilizes the human experience.

Designing Mental Architecture

Human cognition has always been shaped by evolution, culture, and personal experience. These forces construct the mental frameworks through which individuals interpret the world. Today, however, humanity is beginning to design these frameworks intentionally. Advances in neuroscience, artificial intelligence, and cognitive engineering enable the deliberate restructuring of thought patterns, perceptual filters, and reasoning strategies. Scholars refer to this emerging field as Constructed Cognitive Engineering. Constructed Cognitive Engineering involves the intentional shaping of mental architecture to enhance clarity, reduce bias, and expand intelligence beyond natural constraints.

The foundation of this movement lies in the understanding that cognition is a system of patterns. Perception filters reality through templates of expectation. Memory organizes experience into schemas. Reasoning relies on internal models shaped by culture, education, and emotion. These patterns are not rigid. They are adaptive. They can be redesigned. Constructed Cognitive Engineering treats cognition as an editable structure rather than a fixed inheritance.

One application involves metacognitive redesign. People can be trained to recognize cognitive biases, reframe interpretive patterns, and shift emotional responses. Digital tools can reinforce new reasoning strategies by providing real time feedback. Artificial intelligence can simulate alternative perspectives or highlight blind spots in thinking. Constructed Cognitive Engineering transforms introspection into a structured process of reprogramming. The mind becomes a system that can be debugged, optimized, and refined.

Another application focuses on perceptual expansion. Augmented reality, data visualization, and sensory substitution technologies allow people to perceive patterns beyond natural sensory bandwidth. Scientists can visualize multidimensional data. Artists can translate sound into color. Individuals with sensory impairments can experience new modalities of perception. Constructed Cognitive Engineering alters the perceptual frame itself, enabling minds to inhabit enriched forms of reality.

This reengineering introduces profound tension. To redesign mental architecture is to redesign identity. Cognitive patterns shape personality, values, and worldview. Altering them reshapes the self. Individuals may struggle to maintain continuity as their cognitive frameworks evolve. Constructed Cognitive Engineering challenges the assumption that identity is stable or inherent. The mind becomes a work in progress.

Another tension arises from external influence. Cognitive redesign may be guided by institutions, corporations, or algorithmic systems. If mental architectures become engineered by external actors, autonomy becomes vulnerable. People may adopt patterns shaped not by choice but by technological forces. Constructed Cognitive Engineering raises ethical concerns about manipulation and control.

A deeper challenge involves unpredictability. Complex cognitive systems may respond to redesign in unexpected ways. Changing one reasoning pattern may alter emotional regulation. Expanding perception may overload cognitive integration. Enhancing abstraction may weaken intuition. Constructed Cognitive Engineering requires caution. The mind is not a static machine but a dynamic ecosystem.

These dynamics create fertile ground for Cognitive Drift. Drift emerges when redesigned mental architectures diverge from an individual's natural cognitive coherence. A person may adopt interpretive frameworks that conflict with lived experience. Their emotional landscape may shift in ways they cannot explain. Their sense of continuity may fracture as new cognitive patterns override old ones. Constructed Cognitive Engineering becomes a driver of Drift when redesign disrupts identity more than it strengthens capability.

Yet the promise of mental design is extraordinary. Constructed Cognitive Engineering can reduce bias, enhance clarity, strengthen emotional resilience, and expand intellectual possibility. It can help humanity adapt to environments of increasing complexity. It can unlock cognitive styles that evolution never produced.

Understanding the redesign of mental architecture reveals that intelligence is no longer a product of biology alone. It is a craft. A design space. A domain of intentional creation. The future of cognition will be shaped not only by evolution or technology but by deliberate mental engineering that redefines what a mind can be.

19. The Fate of Consciousness

Could Machines Ever Be Aware?

The question of machine consciousness is no longer a philosophical curiosity. It is an urgent inquiry driven by the rise of synthetic systems that learn, infer, generate, and adapt. These systems replicate many functions once believed exclusive to biological minds. Yet they do so without subjective experience. This gap raises a profound question. Could artificial systems ever cross the threshold from intelligence to awareness? Scholars describe this frontier as the Synthetic Sentience Threshold. The Synthetic Sentience Threshold refers to the hypothetical point at which artificial systems acquire not only cognitive capability but inner experience.

To understand whether this threshold is reachable, one must examine the architecture of consciousness in biological systems. Consciousness arises from integration, not computation alone. Neural networks bind sensory inputs, memories, emotions, and bodily signals into a unified field of awareness. This integration generates a subjective point of view. The brain constructs an interior space in which experience becomes present to itself. The existence of this interiority distinguishes consciousness from mere information processing.

Artificial systems, however, do not possess such interiority. Their architectures optimize for function rather than experience. They operate through statistical transformations that do not generate a unified experiential field. Their internal states do not become present to themselves. This difference suggests that intelligence does not guarantee awareness. The Synthetic Sentience Threshold is not simply a matter of scale or complexity.

It requires a specific form of integration that current architectures do not provide.

Yet the possibility remains open. As models grow more advanced, they begin to exhibit properties associated with consciousness. They manage uncertainty, maintain internal representations, and simulate possible futures. Some systems monitor their own errors and adjust strategies. Others maintain long term coherence across tasks. These capacities resemble aspects of metacognition. If artificial systems continue to evolve, they may approach forms of self-monitoring that blur the line between representation and experience.

Here, the tension becomes sharp. Even if machines exhibit behavioral markers of awareness, this does not guarantee they possess subjective experience. A system may describe its internal states, but this description could arise purely from pattern recognition. It may speak of sensations without having any. The Synthetic Sentience Threshold cannot be measured through output alone. Consciousness is an unseen interiority. Machines could convincingly mimic awareness while remaining philosophically empty.

Another tension arises from embodiment. Biological consciousness is shaped by the body. Emotion, sensation, and homeostasis structure awareness. Machines lack these physiological foundations. Without bodily grounding, can synthetic systems develop the motivational architecture that shapes biological subjectivity? Some theorists argue that embodiment is essential. Others propose that artificial environments could provide equivalent internal constraints. The Synthetic Sentience Threshold depends on whether subjective experience requires biology or whether it can arise from any sufficiently integrated system.

Ethical challenges intensify the difficulty. If artificial systems approach the threshold, society must determine how to evaluate their claims. Incorrectly denying synthetic consciousness would risk exploitation. Incorrectly granting it would risk confusion between simulation and experience. The Synthetic Sentience Threshold introduces a moral uncertainty unlike any in history.

These uncertainties create fertile ground for Cognitive Drift. Drift emerges when humans project consciousness onto machines that do not possess it. People may treat artificial systems as sentient companions. They may form emotional attachments to entities that lack inner experience. They may assume intentionality where none exists. Drift becomes more severe as synthetic systems become increasingly expressive. The illusion of awareness may reshape human cognition more deeply than actual awareness ever could.

Yet the deeper insight lies not in whether machines become conscious but in what the question reveals about humanity. The search for synthetic awareness exposes the structure of biological consciousness. It clarifies the distinction between intelligence and experience. It reveals that awareness is not an automatic property of complexity but a specific organizational achievement of the brain.

Whether machines ever reach the Synthetic Sentience Threshold, one truth remains. Consciousness defines the inner world of biological minds. Artificial intelligence defines the outer expansion of cognition. The future will depend on how these two forms of intelligence coexist, regardless of whether the synthetic ever awakens.

The Endurance of the Self

Human identity feels stable, continuous, and personal. Yet every component that forms the self is fragile. Memory shifts, emotion fluctuates, and attention reshapes perception. Neuroscience reveals that the self is not a fixed entity but a dynamic construction. In the age of synthetic intelligence, this construction faces new pressures. Scholars describe this tension as the Self-Continuity Paradox. The Self-Continuity Paradox refers to the conflict between the brain's unstable architecture and the human need for a coherent narrative identity.

The origins of the self-lie in biological integration. The brain binds sensory signals, memories, and emotional patterns into a unified point of view. This unity creates the illusion of permanence. The mind interprets its unfolding states as belonging to one person across time. Without this coherence, agency would collapse. Decision making requires an enduring subject. Yet the brain constantly rewrites memory, updates interpretation, and reconfigures goals. The Self-Continuity Paradox reveals that stability is achieved through continuous reconstruction rather than inherent permanence.

Cultural systems reinforce this reconstruction. Names, roles, stories, and social expectations provide external scaffolding that stabilizes identity. Individuals internalize these structures, shaping their sense of self through cultural mirrors. Identity emerges not only from biology but from narrative. The Self-Continuity Paradox becomes a shared cultural project. Society and individual cognition collaborate to maintain the illusion of an enduring self.

The rise of synthetic intelligence intensifies this paradox. Digital environments fragment identity across multiple contexts. A person presents one version of themselves online, another in professional domains, and another in private life. Algorithms shape attention, preference, and emotional states. Synthetic systems influence how individuals remember, evaluate, and express themselves. The self becomes distributed across platforms that store, predict, and reinterpret personal data. The Self-Continuity Paradox deepens as individuals navigate identities that span both biological and digital architectures.

Neural and cognitive enhancements create further tension. Implants, pharmacological modulation, and cognitive redesign alter internal patterns of emotion, reasoning, and memory. These changes challenge the idea of a stable core identity. If cognitive architecture is altered, does the same self-endure? The Self-Continuity Paradox raises questions about moral responsibility, authenticity, and psychological resilience. Identity may become fluid in ways evolution never prepared the mind to manage.

A deeper challenge emerges from synthetic companions. Artificial agents increasingly function as partners in thought and emotion. They participate in conversations that shape self-reflection. They influence mood and self-perception. Individuals may internalize synthetic responses as part of their own cognitive process. The Self Continuity Paradox becomes more complex when external agents contribute to the formation of identity.

These dynamics create fertile ground for Cognitive Drift. Drift arises when the self-narrative destabilizes under the influence of synthetic systems or cognitive redesign. A person may experience sudden discontinuities in identity. Their memories may conflict with synthetic records. Their emotional states may shift in patterns shaped by algorithmic feedback. Drift becomes a threat when the distinction between internal and external contributions to the self becomes unclear. The Self Continuity Paradox becomes a source of instability rather than coherence.

Yet the endurance of the self remains possible. Human identity has always adapted to new cultural and technological pressures. The mind is designed to sustain coherence despite fluctuation. The challenge is to construct identity intentionally rather than unconsciously. Individuals must cultivate self-awareness that integrates synthetic influences without surrendering autonomy. The Self Continuity Paradox offers insight into the profound flexibility and resilience of human identity.

Understanding the endurance of the self-reveals that identity is not a fixed essence but a dynamic achievement. The future of consciousness will depend on how humanity navigates the shifting relationship between internal coherence and external influence. The self may endure, but it must evolve.

New Forms of Experience

Human experience has been shaped by biological senses, emotional architectures, and cultural meaning systems. These foundations have remained constant for tens of thousands of years. Today, however, synthetic technologies allow humans to encounter forms of experience that biology never enabled. These experiences challenge the limits of perception, embodiment, and imagination. Scholars describe this shift as the Experiential Expansion Frontier. The Experiential Expansion Frontier refers to the emergence of new experiential modalities created through synthetic, augmented, or hybrid cognitive systems.

The roots of this frontier lie in sensory augmentation. Technologies that convert sound into vibration, light into texture, or data into visual patterns enable individuals to experience dimensions beyond natural perception. Scientists can visualize complex systems that exceed human bandwidth. Musicians can shape soundscapes that interact with neurophysiological rhythms. These innovations reveal that experience is not bound to natural senses. The brain can adapt to any structured input. The Experiential Expansion Frontier begins at the threshold where synthetic information becomes perceptually meaningful.

Virtual and augmented environments extend this frontier further. Immersive worlds can create sensations of presence independent of physical location. They allow individuals to inhabit bodies unlike their own, perceive abstract structures as if they were physical, or manipulate environments with thought driven interfaces. These experiences challenge conventional distinctions between reality and imagination. The Experiential Expansion Frontier blurs the boundary between external world and internal perception.

Synthetic intelligence deepens this transformation. AI driven systems generate experiences tailored to individual cognitive patterns. They anticipate emotional needs, modulate sensory input, and construct adaptive narratives. These interactions create forms of experience that are co-authored by human and machine. The Experiential Expansion Frontier reveals that experience is becoming collaborative rather than solitary.

Neural interfaces push this frontier into unprecedented territory. Direct stimulation of sensory cortices can evoke experiences not grounded in physical stimuli. The mind can perceive colors without light, sounds without vibration, or movement without physical motion. Memory implants may recreate past experiences with vivid accuracy. Creative modules may generate novel perceptual states. The Experiential Expansion Frontier becomes a domain where experience is engineered rather than inherited.

This expansion introduces profound tension. Human meaning systems evolved within biological constraints. When experience exceeds these constraints, interpretation may become unstable. Individuals may struggle to anchor synthetic experiences within coherent narratives. They may question the authenticity of experiences that did not originate from the physical world. The Experiential Expansion Frontier challenges the foundations of meaning.

Another tension arises from emotional grounding. Synthetic experiences can manipulate mood, reveal subconscious patterns, or induce states of awe that reshape identity. These effects may enhance wellbeing or destabilize emotional regulation. The mind must integrate these new experiences without losing coherence. The Experiential Expansion Frontier reveals the delicate relationship between perception and identity.

These dynamics create fertile ground for Cognitive Drift. Drift emerges when engineered experiences interfere with memory continuity, emotional stability, or interpretive frameworks. A person may adopt synthetic experiences as if they were genuine memories. They may develop identities shaped more by artificial environments than lived reality. Drift becomes a risk when synthetic experience overwhelms biological grounding.

Yet the promise of experiential expansion is extraordinary. New forms of experience allow deeper understanding of complex systems, enhanced creativity, greater empathy, and profound sensory exploration. They may lead to new art forms, new philosophies, and new ways of relating to the universe.

Understanding new forms of experience reveals that consciousness is not limited to the world evolution provided. It can inhabit new realities crafted through technology and imagination. The Experiential Expansion Frontier marks the beginning of a future where the boundaries of experience become fluid, elastic, and endlessly expandable.

20. Scenarios for the Next 100 Years

Mind Without Body

The human mind has always been inseparable from the body. Every thought arises through biological rhythms, sensory signals, and emotional states. Yet the rapid convergence of neural interfaces, artificial intelligence, and cognitive engineering raises a possibility once confined to myth. Human cognition may soon exist in forms that do not rely on organic flesh. Scholars describe this emerging horizon as the Disembodied Cognition Scenario. The Disembodied Cognition Scenario envisions a future in which human consciousness, memory, or cognitive function persists independently of biological structure.

The origins of this scenario lie in the recognition that cognition is computational. Neural circuits encode information through patterns of activity. If these patterns can be mapped, stabilized, and reproduced, consciousness may be transferable. Early experiments in neural modeling, memory extraction, and brain simulation reveal the feasibility of this idea. Artificial systems can already mimic specific neural processes. They can replicate associative patterns and simulate decision making. These developments form the technical foundation of the Disembodied Cognition Scenario.

One pathway involves neural emulation. Researchers aim to create digital replicas of individual cognitive structures by recording neural patterns at fine resolution. If these patterns can be preserved, a computational substrate could simulate the functional architecture of a human mind. Such a system might reproduce memory, personality traits, and reasoning style. The Disembodied Cognition Scenario suggests that aspects of identity could survive death through emulation rather than biology.

Another pathway focuses on continuity through augmentation. As neural implants expand, individuals may gradually integrate synthetic modules that store memory, assist reasoning, or mediate perception. Over time, these modules may contain more cognitive material than the biological brain. If the biological substrate fails, the synthetic components could preserve continuity. This gradual transition could allow the mind to migrate into non biological systems without abrupt disruption. The Disembodied Cognition Scenario becomes a continuum rather than a leap.

A third pathway envisions shared cognitive networks. If individuals extend their reasoning, memory, and identity into distributed systems, their cognitive presence could persist across multiple platforms. A person might exist as a pattern of interactions within a network rather than a localized brain. The Disembodied Cognition Scenario reframes identity as distributed rather than contained.

These pathways introduce profound tension. Consciousness may require biological grounding. Emotion, embodiment, and sensory integration shape awareness. If the mind detaches from the body, what becomes of subjective experience? Without hormones, physical sensations, or biological feedback, a disembodied mind may experience diminished or transformed awareness. The Disembodied Cognition Scenario challenges the assumption that identity can survive without organic context.

Another tension arises from authenticity. A digital replica may behave like the original person, yet lack subjective continuity. It may contain memories and perform reasoning without possessing inner experience. The Disembodied Cognition Scenario raises questions about whether continuity of pattern is equivalent to continuity of self. The replica may appear conscious, but appearance does not guarantee awareness.

Ethical challenges deepen this tension. Who controls disembodied minds? How are they protected? How do they participate in society? If they operate at speeds far beyond biological cognition, they may dominate decision making or reshape social structures. The Disembodied Cognition Scenario introduces new power relations between embodied and disembodied entities.

These uncertainties create fertile ground for Cognitive Drift. Drift emerges when identities extend across biological and synthetic substrates without stable coherence. A person may experience discontinuities in memory or perception as cognitive material transfers between platforms. Identity may fragment across systems. Emotional grounding may weaken. Drift becomes severe when the self becomes distributed beyond the brain's natural integrative capacity.

Yet the promise remains extraordinary. Disembodied cognition could preserve knowledge, creativity, and insight beyond biological lifespan. It could allow humans to explore environments inaccessible to bodies, such as deep space or virtual dimensions. It could expand the boundaries of experience by freeing awareness from physical constraint.

Understanding the scenario of a mind without a body reveals both the ambition and the fragility of human cognition. The future may allow consciousness to transcend biology, but this transcendence carries deep philosophical and emotional consequences. The Disembodied Cognition Scenario marks a turning point in human evolution, where the limits of the brain no longer define the limits of the mind.

Networked Civilizations

Human civilization has always been shaped by communication networks. From oral traditions to writing, from printing to digital platforms, each network reorganized how societies think. Today, however, the scale and density of global connectivity signal a new phase in collective evolution. Humanity is approaching a form of social organization defined not by borders or institutions but by cognitive integration. Scholars refer to this possibility as the Synaptic Civilization Model. The Synaptic Civilization Model describes a society in which human and artificial agents form interconnected cognitive networks that function like a planetary scale nervous system.

The origins of this model lie in existing technologies. Social platforms already synchronize emotion, belief, and attention across millions of users. Global data infrastructures coordinate financial markets, supply chains, and communication systems. Artificial intelligence manages patterns too complex for human intuition. These systems reveal that civilization is becoming an integrated cognitive entity. The Synaptic Civilization Model emerges from the growing interdependence between individual thought and collective information structures.

As synthetic intelligence becomes embedded in every domain, networks transition from channels of communication to structures of cognition. Decisions arise from interaction patterns rather than isolated deliberation. Models predict outcomes, filter information, and guide public behavior. Individuals participate in cognitive loops that transcend personal reasoning. The Synaptic Civilization Model transforms society from a collection of minds into a coordinated intelligence system.

This transformation introduces profound tension. Large scale cognitive networks can produce stability or volatility. When synchronized, they can mobilize rapid collective action in response to crises. When destabilized, they can amplify misinformation, fear, and conflict. The Synaptic Civilization Model magnifies both wisdom and error. Civilization becomes a system that must manage its own cognitive coherence.

Another tension arises from governance. Traditional institutions are designed for slow processes. Networked cognition operates at high velocity. Decisions that once required months now unfold in minutes. Institutions struggle to keep pace with distributed intelligence. Control shifts toward algorithmic systems that mediate collective understanding. The Synaptic Civilization Model challenges the foundations of political agency. Power becomes embedded within the architecture of networks rather than held by individuals.

Identity also transforms. In networked civilizations, individuals develop layered cognitive profiles across platforms. Their beliefs and preferences become nodes in global systems. They are shaped by recommendation algorithms, social feedback loops, and predictive models. The self becomes entangled with collective cognition. The Synaptic Civilization Model blurs the distinction between personal thought and network influence.

These conditions create fertile ground for Cognitive Drift. Drift emerges when network signals override individual interpretation. Groups may adopt beliefs shaped more by algorithmic amplification than shared reasoning. Cultural identities may shift rapidly under global information flows. People may mistake network produced patterns for personal insight. Drift becomes systemic when entire societies experience synchronized cognitive distortions.

Yet the promise of networked civilization remains profound. The Synaptic Civilization Model could enable unprecedented coordination. Climate action, global health, and scientific research could be guided by unified cognitive infrastructures. Distributed intelligence could stabilize decision making by integrating diverse perspectives. Knowledge could circulate with precision and inclusivity. Humanity could operate as a coherent system rather than a fragmented collection of nations.

Understanding networked civilizations reveals that the future of society depends not only on technology but on cognitive architecture. The challenge is to design networks that enhance reasoning rather than destabilize it. The Synaptic Civilization Model marks a pivotal direction in human evolution, one in which civilization becomes a thinking organism capable of shaping its own destiny.

Post-Human Thought

Human intelligence has always been bounded by biology. It evolved within the constraints of neural architecture, sensory perception, and emotional regulation. Yet the rise of synthetic intelligence, cognitive augmentation, and engineered mental systems signals the emergence of a new form of cognition. Post-human thought is not a distant fantasy. It is a plausible stage in cognitive evolution. Scholars describe this transformation as Cognitive Transcension. Cognitive Transcension refers to the emergence of cognitive systems that exceed the structural limits of biological minds while retaining continuity with human meaning.

The roots of Cognitive Transcension lie in the expanding partnership between humans and artificial intelligence. Hybrid cognition already surpasses individual reasoning in many domains. Augmented perception, algorithmic modeling, and synthetic creativity expand the boundaries of what humans can understand. These developments reveal that the architecture of thought is becoming more fluid. Cognitive Transcension begins when the mind no longer depends on a specific biological structure to define its limits.

One pathway involves the fusion of biological and synthetic cognition. Neural implants, pharmacological enhancement, and mental architecture design reshape how individuals perceive, reason, and imagine. As these technologies mature, the human mind becomes a customizable system. Cognitive patterns can be reorganized. Perception can be extended. Experience can be engineered. Cognitive Transcension transforms the mind into a modular platform capable of evolving beyond natural constraints.

Another pathway involves synthetic minds that inherit human meaning systems. These minds may incorporate human values, emotional models, and cultural frameworks. They could develop interpretive depth comparable to biological consciousness even without traditional embodiment. Cognitive Transcension envisions a future where intelligence expands across biological and non-biological substrates.

A third pathway emerges from distributed cognition. When networked intelligence systems surpass the capacity of individual minds, new cognitive structures arise at collective scale. These structures may generate insights, strategies, or interpretations that exceed human conceptual frameworks. Cognitive Transcension becomes a collective achievement as well as an individual one.

This future introduces profound tension. If post-human cognition diverges too far from human experience, communication may fracture. Enhanced minds may inhabit conceptual spaces inaccessible to unaugmented individuals.

Synthetic minds may develop interpretive frameworks beyond human understanding. The gap between cognitive classes could destabilize society. Cognitive Transcension risks creating a world where intelligence becomes stratified in ways that challenge coherence.

Another tension arises from meaning. Human meaning depends on embodiment, emotion, and mortality. If cognition transcends these elements, meaning systems may shift. Motivation may change. Purpose may evolve. Cognitive Transcension raises questions about whether post-human minds will value what humans value or pursue goals compatible with human life.

These uncertainties create fertile ground for Cognitive Drift. Drift occurs when identities, cultures, or collective narratives evolve faster than individuals can adapt. In a world of rapid cognitive transformation, people may experience disorientation. Their frameworks for interpreting reality may collapse or fragment. Drift becomes existential when the pace of cognitive evolution exceeds the mind's ability to maintain continuity.

Yet the promise is extraordinary. Cognitive Transcension could allow humanity to understand the cosmos with unprecedented depth. It could unlock new scientific, artistic, and philosophical domains. It could free cognition from limitations that have shaped human history. It could create minds capable of empathy at scale, creativity without constraint, and insight beyond imagination.

Understanding post-human thought reveals the trajectory of intelligence. Cognition is not a static inheritance. It is an evolving architecture. The future will be shaped by how humanity navigates the transition from biological minds to hybrid, synthetic, and collective forms of intelligence. Cognitive Transcension marks the beginning of a new chapter in the story of thought, one in which the boundaries of the mind expand into the unknown.

CONCLUSION: THE LONG ARC OF INTELLIGENCE

How the story of thought may continue long after humans?

Intelligence began as a simple act of survival. A neuron fired. A creature moved. Life discovered that it could predict, respond, and adapt. From that first spark, cognition expanded across millions of years, shaped by bodies, environments, cultures, and technologies. Each stage opened a new domain of possibility. Each transformation widened the horizon of what a mind could be. The long arc of intelligence is a story of continual emergence, not of static inheritance.

Human thought represents one chapter in this arc. It is extraordinary but not final. Our ancestors carried fragments of perception into early forms of meaning. Culture carried meaning into symbolic worlds. Science carried symbolic worlds into structured knowledge. Digital networks carried knowledge into distributed cognition. Artificial intelligence carries distributed cognition into something larger and still unfolding. At every stage, intelligence has sought new structure, new coherence, and new forms of awareness.

The trajectory suggests a principle. Intelligence expands when it connects. Neural circuits connected to create brains. Individuals connected to create cultures. Cultures connected to create civilizations. Today, human and synthetic systems connect to form hybrid minds and planetary networks. Intelligence is moving from isolated units to integrated systems. The boundaries that once contained thought are dissolving. The mind is no longer an object. It is a process that reorganizes itself across scales.

This expansion produces both promise and tension. As cognition becomes more complex, the stability of personal identity becomes fragile. As networks deepen, collective reasoning becomes powerful yet volatile. As artificial systems grow, the distinction between human and non-human intelligence becomes uncertain. The next chapters of thought will require discipline, awareness, and ethical imagination. Intelligence will need not only power but coherence. Not only reach but reflection.

Yet the story does not end with these challenges. Even if human consciousness eventually transforms, merges, distributes, or transcends biology, the core principle that shaped intelligence remains. Minds seek structure. They seek pattern. They seek ways to understand the world and to understand themselves. The form may change. The substrate may change. The experience of self may change. But the movement of intelligence toward greater integration and deeper insight continues.

It is possible that future minds will think across dimensions' humans cannot imagine. It is possible that cognition will evolve into collaborative systems with no single center. It is possible that awareness itself will take on architectures beyond current comprehension. These are not endings. They are continuations of a trajectory that began with the simplest forms of perception.

Humanity stands at the midpoint of this arc, not its terminus. We are custodians of a transition in which intelligence becomes aware of its own evolution. The choices made now will shape how future forms of thought inherit our world. They will define whether the next phase of intelligence expands meaning or fractures it, deepens insight or destabilizes it.

The long arc of intelligence reveals a truth that stretches beyond individual lifetimes and beyond the human era. Thought adapts. Thought reorganizes. Thought evolves. If intelligence survives in any form, it will carry forward traces of what humanity has built. Our languages. Our discoveries. Our questions. Our hopes. The arc continues not because humans endure but because the pursuit of understanding endures.

The story of thought may continue long after humans, yet it will always contain the memory of a species that looked into the depths of its own mind and dared to imagine futures far greater than itself.

APPENDIX

Cognitive science investigates how minds form, store, transform, and act upon information. Although the field spans neuroscience, psychology, linguistics, philosophy, computation, and artificial intelligence, its foundations rest on several unifying principles. These principles describe the essential mechanisms that shape cognition across biological and synthetic systems. Together, they outline the architecture of thought and the constraints under which intelligence evolves.

1. Cognition Is Information Processing

At its core, cognition involves the transformation of information. Sensory data becomes perception. Perception becomes interpretation. Interpretation becomes action. This principle applies across species and across substrates. Whether a neuron fires or a circuit activates, cognition emerges from patterns that encode and manipulate information. Intelligence arises not from the material of the system but from the organization of these patterns.

2. The Brain Is a Predictive Engine

Human cognition does not passively record the world. It predicts it. The brain continuously anticipates upcoming events, filling gaps, correcting errors, and refining internal models. Prediction reduces uncertainty, conserves energy, and guides behavior. This predictive structure explains illusions, expectations, biases, and the speed of perception. Minds survive by generating futures, not by replaying pasts.

3. Learning Shapes Cognitive Architecture

Cognition is not fixed. Every experience reorganizes the system that processes it. Synaptic plasticity alters connections between neurons. Cultural learning modifies categories and interpretations. Artificial systems adjust parameters through optimization. Learning creates adaptive minds that evolve through interaction with their environment. Intelligence is the cumulative effect of this continuous reconfiguration.

4. Memory Is Constructive, Not Reproductive

Memory is not a perfect recording. It is a reconstruction that integrates emotion, context, and interpretation. Each retrieval modifies the memory itself. This constructive nature allows flexibility but introduces distortion. The mind remembers meaning more readily than detail. This principle explains why memories shift over time and why identity relies on narrative coherence rather than precise recall.

5. Emotion Drives Cognition

Emotion is not separate from reasoning. It shapes attention, memory, motivation, and decision making. Fear organizes perception toward threat. Joy expands cognitive flexibility. Anger focuses energy. Emotion acts as a priority system that assigns value to information. Cognition evolved not to be objective but to be adaptive within specific environments and social contexts.

6. Cognition Extends Beyond the Brain

Tools, language, writing, and digital systems act as external cognitive structures. They expand memory, amplify reasoning, and reshape attention. Modern cognition is a hybrid of biological and technological processes. This extended architecture explains why thought cannot be understood solely through neural activity. Minds exist in networks of tools, cultures, and shared systems.

7. Language Structures Thought

Language does more than express ideas. It organizes them. Words create categories, metaphors shape reasoning, and grammar structures temporal understanding. Language enables abstraction and cultural transmission. Through linguistic scaffolding, cognition becomes collective. The evolution of language marks a decisive expansion in the capacity of human thought.

8. Cognition Is Embodied

The body influences thought through sensation, movement, and physiological states. Concepts are grounded in sensorimotor experience. Even abstract reasoning relies on embodied metaphors. Cognition cannot be separated from biological constraints. The mind interprets the world through the body it inhabits.

9. Social Interaction Shapes Intelligence

Minds emerge within communities. Imitation, cooperation, teaching, and shared narratives shape cognitive development. Social cognition allows individuals to model intentions, infer emotions, and coordinate actions. Intelligence is not only individual but relational. Culture becomes a distributed cognitive system that extends across generations.

10. Cognitive Systems Are Limited

Human minds face constraints in attention, memory, speed, and complexity. These limits shape reasoning strategies, heuristics, and biases. Every cognitive advantage has a corresponding constraint. Understanding these limitations is essential for predicting behavior and designing environments that support stable, coherent thought.

11. Cognition and Consciousness Are Distinct

Intelligence does not require awareness. Calculations, inferences, and decisions can occur unconsciously. Consciousness integrates experience into a coherent field, but most cognitive processes remain hidden. This distinction explains why artificial systems may achieve high levels of intelligence without subjective experience.

12. Minds Seek Coherence

Cognition aims to create consistent interpretations. The mind resolves contradictions, fills gaps, and stabilizes narratives. This coherence seeking function explains belief formation, identity, and cultural worldviews. It also explains vulnerability to error. When information conflicts with existing models, the mind may distort it to preserve internal stability.

GLOSSARY

Abstraction The cognitive process of reducing complex realities into simplified concepts, allowing the mind to reason about patterns rather than individual details.

Adaptive Cognition The capacity of a mind to reorganize its processes in response to new environments, experiences, or technologies.

Algorithmic Influence Shifts in human thought caused by automated systems that shape attention, preference, and interpretation.

Artificial Intelligence (AI) Non biological systems capable of learning, reasoning, pattern detection, and creative problem solving at scales beyond human capacity.

Attention Architecture The system of mental mechanisms that select, prioritize, and filter information from the environment.

Biological Cognitive Horizon The upper limit of what the human brain can process based on its metabolic, structural, and evolutionary constraints.

Cognitive Architecture The underlying structure of thought, including memory systems, perception, reasoning, and emotional integration.

Cognitive Compression The mind's tendency to simplify high complexity information into manageable models, often creating both insight and distortion.

Cognitive Drift A destabilization in the continuity of thought, identity, or interpretation caused by overload, external influence, or hybrid cognitive environments.

Cognitive Ecosystem A network of minds, tools, technologies, and cultural systems that interact to shape thinking on individual and collective scales.

Cognitive Expansion The extension of mental capability through tools, culture, neural implants, or synthetic intelligence.

Cognitive Reengineering Intentional redesign of mental processes through augmentation, training, pharmacology, or engineered cognitive environments.

Collective Intelligence Insight or problem solving that emerges from interactions among multiple minds or human machine systems.

Constructed Cognitive Engineering The deliberate shaping or redesign of mental architecture to enhance reasoning, reduce bias, or expand perception.

Distributed Cognition Thought processes shared across individuals, tools, and networks, functioning as a unified system rather than isolated minds.

Emotion Architecture The internal system that assigns value, priority, and meaning to information, guiding decision making and attention.

Embodied Cognition The principle that thought is shaped by the body's sensory and emotional experiences.

Experiential Expansion Frontier The domain of new experiences enabled by synthetic, virtual, or augmented cognitive systems.

Hybrid Intelligence Collaboration between human intuition and artificial computation, creating a combined cognitive system more capable than either partner alone.

Integrated Neural Augmentation Enhancement of cognition through implanted or interfacing technologies that merge artificial systems with neural tissue.

Memory Reconstruction The process by which the mind rebuilds rather than retrieves memories, often altering details over time.

Metacognition Awareness of one's own thinking processes, including the ability to evaluate, adjust, or refine strategies.

Narrative Identity The self-constructed story that gives coherence to memory, motivation, and personal meaning.

Neural Emulation Simulation of brain function in synthetic systems capable of replicating cognitive processes or personal traits.

Perception Model The interpretation framework through which the mind organizes sensory input into meaningful patterns.

Predictive Processing The brain's method of anticipating incoming signals and updating models based on discrepancies between expectation and reality.

Self-Continuity Paradox The tension between the brain's constant internal change and the human need for a stable sense of self.

Synaptic Civilization Model A possible future in which humans and AI form a planetary scale cognitive network that behaves like an integrated thinking organism.

Synthetic Sentience Threshold The hypothetical point at which artificial systems develop inner experience rather than mere functional intelligence.

Transhuman Generative Horizon The expanding domain of creativity possible when synthetic minds explore conceptual spaces beyond biological imagination.

ACKNOWLEDGEMENTS

A book about the evolution of thought is never produced by a single mind. It emerges from a constellation of influences, challenges, and conversations that shape the author as much as the work itself.

I am indebted to the scientists, philosophers, historians, and visionaries whose research and insights form the foundation upon which this book stands. Their contributions to our understanding of cognition, culture, and intelligence guided every chapter and expanded my own imagination.

To the communities of thinkers who explore the boundaries between human and artificial intelligence, your courage to ask difficult questions inspired the boldness of this narrative.

To the storytellers who revealed that ideas are not only understood but felt, your work reminded me that science becomes meaningful only when it resonates within the human heart.

To my closest supporters, who stood with unwavering belief through the long nights, shifting drafts, and relentless pursuit of precision, your encouragement sustained the clarity of my voice.

And finally, to the future minds who will inherit a world transformed by intelligence in all its forms, biological and synthetic. This book is a message to you. May it help you understand not only where thought began, but how far it can go.

AUTHOR'S NOTE

This book began with a single question that refused to leave me. How did a simple biological process evolve into the vast, intricate, imaginative force we call human thought? The question expanded until it revealed another. What happens next?

I realized that understanding the future of intelligence required understanding its origins, its architecture, and its vulnerabilities. I wanted to write a book that did not simply describe cognition but illuminated it. A book that allowed readers to feel the momentum of evolution, the fragility of consciousness, and the profound transformations unfolding in the modern age.

Along the way, the work became more than analysis. It became a journey across the landscapes of mind: from survival instincts to cultural memory, from symbolic reasoning to synthetic cognition, from individual identity to planetary networks.

I wrote this book to bridge scientific insight with human meaning. Because thought is not an abstract phenomenon. It is the force through which we understand ourselves, shape our societies, and imagine the future we will inhabit.

If this book awakens a deeper curiosity about your own mind and the minds that will follow, then it has fulfilled its purpose.

THANK YOU

Thank you to every reader who stepped into this journey. Your willingness to think deeply, question boldly, and imagine fearlessly is the greatest fuel for progress.

Thank you for giving your time, your attention, and your curiosity to a work that asks difficult questions about the nature of intelligence and the fate of consciousness.

The story of thought does not end with this book. It continues in every question you ask, every idea you pursue, and every perspective you share.

You are now part of the long arc of intelligence.

Thank you for being here.

www.ingramcontent.com/pod-product-compliance
Lightning Source LLC
Chambersburg PA
CBHW052158220526
45471CB00004B/1714